COLLECTED POEMS
1930–1986

BY THE SAME AUTHOR

Poetry

A Bravery of Earth
Reading the Spirit
Song and Idea
Poems New and Selected
Burr Oaks
Brotherhood of Men
An Herb Basket
Selected Poems
Undercliff: Poems 1946–1953
Great Praises
Collected Poems 1930–1960
Collected Verse Plays
The Quarry
Selected Poems 1930–1965
Thirty One Sonnets
Shifts of Being
Fields of Grace
Poems to Poets
To Eberhart from Ginsberg
Collected Poems 1930–1976
Ways of Light
Survivors
Four Poems
New Hampshire/Nine Poems
Chocorua
Florida Poems
The Long Reach

Criticism

Of Poetry and Poets

Anthology

War and the Poet (with Selden Rodman)

Richard Eberhart

COLLECTED POEMS

POEMS

1930–1986

NEW YORK   OXFORD

OXFORD UNIVERSITY PRESS

1988

Oxford University Press
Oxford   New York   Toronto
Delhi   Bombay   Calcutta   Madras   Karachi
Petaling Jaya   Singapore   Hong Kong   Tokyo
Nairobi   Dar es Salaam   Cape Town
Melbourne   Auckland

and associated companies in
Beirut   Berlin   Ibadan   Nicosia

Published by Oxford University Press, Inc.,
200 Madison Avenue, New York, New York 10016

Oxford is a registered trademark of Oxford University Press

Library of Congress Cataloging-in-Publication Data
Eberhart, Richard, 1904–
Collected poems, 1930–1986.
1. Title.
PS3509.B456A6   1987   811'.52   87-15236
ISBN 0-19-504055-4

Previously published poems in this book first appeared in the following collec-
tions: A Bravery of Earth, Reading the Spirit, Song and Idea, Poems New and
Selected, Burr Oaks, Selected Poems, Under Cliff, Great Praises, Collected Poems
1930–1960, The Quarry, Selected Poems 1930–1965, Shifts of Being, Fields of
Grace, Ways of Light, The Long Reach, Four Poems, Florida Poems, Breadloaf
Anthology.

1 3 5 7 9 8 6 4 2

Printed in the United States of America
on acid-free paper

ACKNOWLEDGMENTS

I am grateful to the editors of the following publications in which some of the poems in this book first appeared: Agenda, The American Poetry Review, Antaeus, Beloit Poetry Journal, **Best Poems of 1971, Borestone Mountain Poetry Awards 1972,** Chelsea Retrospectives, Chicago Tribune, Concerning Poetry, Counter-Measures, The Dartmouth, Darthmouth Alumni Magazine, The Dart, Florida Quarterly, Fragments, The Harvard Advocate, Harvard Magazine, Hellcoal Annual Two, The London Magazine, The London Times Literary Supplement, Michigan Quarterly Review, The Nation, The New York Times, The New York Quarterly, Northern Lights, Periphery, Pomegranate Press, Poetry, Poetry Now, Prairie Schooner, Pulse, Quadrant, Quarterly Review of Literature, Saturday Review, Shaman, Smith College Library Bulletin, The South Florida Poetry Journal, The Southern Review, The Sou'wester, Stand, Stone Drum, The Virginia Quarterly Review, West Hills Review, Yale Literary Magazine.

The following poems first appeared in The New Yorker: "As If You Had Never Been", "Despair", "The Fisher Cat", "Homage to the North", "Suicide Note", "Track", "Emblem", "Slow Boat Ride", and "Coast of Maine".

"Sailing to Buck's Harbor" first appeared in *The Yale Literary Magazine,* Vol. 150, No. 1. Reprinted by permission © 1982 by American Literary Society.

Poems from *The Long Reach* (copyright © 1948, 1953, 1956, 1964, 1967, 1976, 1977, 1978, 1980, 1981, 1982, 1983, 1984 by Richard Eberhart) are reprinted by permission of New Directions Publishing Corporation.

CONTENTS

vi

Contents

Contents

Contents

Contents

Contents

Contents

Contents

Contents

xiv

Contents

Contents

COLLECTED POEMS

1930–1986

Poetry goes down into the darkness of the subconscious and comes up with a morning glory out of the enriched mud and schist of the world. The morning glory could be a lily, or a rose, Chartres Cathedral, or the Parthenon, or John Keats, a quality of perfection dreamed of by mankind.

Barbizon Plaza Hotel, New York, 5 A.M., May 18, 1983.
Read at induction into the American Academy of Arts and Letters
at the annual ceremony of the Academy.

THIS FEVERS ME

This fevers me, this sun on green,
On grass glowing, this young spring.
The secret hallowing is come,
Regenerate sudden incarnation,
Mystery made visible
In growth, yet subtly veiled in all,
Ununderstandable in grass,
In flowers, and in the human heart,
This lyric mortal loveliness,
The earth breathing, and the sun.
The young lambs sport, none udderless.
Rabbits dash beneath the brush.
Crocuses have come; wind flowers
Tremble against quick April.
Violets put on the night's blue,
Primroses wear the pale dawn,
The gold daffodils have stolen
From the sun. New grass leaps up;
Gorse yellows, starred with day;
The willow is a graceful dancer
Poised; the poplar poises too.
The apple takes the seafoam's light,
And the evergreen tree is densely bright.
April, April, when will he
Be gaunt, be old, who is so young?
This fevers me, this sun on green,
On grass glowing, this young spring.

O WILD CHAOS!

O wild Chaos! O sweet Chaos!
Dance to my arms, I would lie with you,
With your last dancing garment off!
O sweet Chaos! O wild Chaos!

I

Let me get you in hilarious weight
With the maddest world's merriest child.
Sweet, sweet Chaos with burningest lips,
And your limbs, they make a goddess unable!
Come to me with your giddiest love,
Wild, wild, sweet, sweet Chaos!

THE BELLS OF A CHINESE TEMPLE

The bells of a Chinese temple sang
A monody in sunned Sabang,
A singing timbreless of tone,
Like water falling on a stone.
And he was glad, glad to be
A sailor resting from the sea
Where Indians and Arabs go,
Most slowly, slow and slow;
Where swart Malayan women walk
Dreamfully along and talk;
Where skirted, bearded, dark-skin Turks
Go hand in hand, and no one works.
China women with bound doll feet
Like marionettes move down the street,
And underneath the arbored trees,
Mild-eyed, squatting in twos or threes,
Chinamen smoke, peaceful and still,
Gazing afar at the palms on the hill,
Pondering Buddha, these Chinamen—
Or counting their hoarded yen.
And there are girls whose smiles are worth
Some subtle Asiatic mirth
Understood and wonderful:
Girls dark, and shy, and beautiful.

He had not known harbor or town
So free from tumult, fret, or frown,

Of all the sunny towns that are
Spun under a southern star
Upon the Oriental South.
Here Love with amorous mellow mouth
Drinks from the chalice of delight
Sun-mulled wine from dawn till night.
Here the ponderer plucks a lute;
The drowsed land is ripe with fruit;
And all man's conquest and man's glory
Is but a story-teller's story,
Incredible and strangely told
Of men far off amazed with gold,
Who bend beneath some heavy plan
To trample down a fellow-man,
Thus harvest wealth, and fame, and power.

This is Sebang; hour on hour
The full day ripens in the sun,
And time has always just begun.
Primeval silence without stir
Holds the earth like gossamer;
For love is slowly blossoming
In quietness too still to sing,
From which all passions green or ripe
Are shadowy blooms of the Immortal Type.

And here he lingers, murmuring
The name of some forgotten king,
Who had a wonder-welling heart:
Richard Ghormley Eberhart.

MAZE

I have a tree in my arm,
There are two hounds in my feet,
The earth can do me no harm
And the lake of my eyes is sweet.

But a fire has burnt the tree down,
I have no blood for the hounds.
Why has the will made me a crown
For a human mind that has bounds?

Who made the tree? Who made fire?
The hounds have gone back to the master.
The earth has killed my desire
That leaped up faster and faster.

It is man did it, man,
Who imagined imagination,
And he did what man can,
He uncreated creation.

There is no tree in my arm,
I have no hounds in my feet,
The earth can soothe me and harm,
And the lake of my eyes is a cheat.

FOR A LAMB

I saw on the slant hill a putrid lamb,
Propped with daisies. The sleep looked deep,
The face nudged in the green pillow
But the guts were out for crows to eat.

Where's the lamb? whose tender plaint
Said all for the mute breezes.
Say he's in the wind somewhere,
Say, there's a lamb in the daisies.

4

CARAVAN OF SILENCE

(For J. L. Sweeney)

Third with silence in a tent
He, unamazed in meditation,
Looked up from Persia, where he sent
Spiraling wings beyond Creation.

Drowsily the one beside him
Laid his elbow on sleep,
Wines of sunned hills denied him
Moods of moods he could not keep;
He slept, with night living around
Like sea on a dark-fathomed plant.
His was a silence without sound
To beetle, grasshopper, or ant.
Camels and he like dunes would sleep
Till the gold ball came over,
Then stray the tedious journey, keep
The mild destiny of the rover:
Hunger and sleep, and the slight
Untroubling dreams, ununderstood,
And never labouring the farsight
Of cities where no cities stood.

Drowsily he breathed. But by him,
Posed, the second sat, gave out
Upon the vast of thoughts that sky him
Silence to make the stars shout.
The audible bloom of night
He heard bell and rebell desire;
His silence trembled the world, that might
Give back and echo and inspire.
He toed warm sand, probed
His fingers through the warm air.
Purpose by the night was robed
Secure. He fought no despair.

There is a green land, and water
Clear as the night is scarved, place
Where wait wife and daughter,
And the opal, unborn face.
The wish is a loud silence
And the hope is a star power
That shake night with innocence
Soft hour on dark hour.
Thus he, the second silence, flung
A bold tone he could recapture.
The cloth of night about him hung,
He took the stigmata of rapture.

But a ghost form was the third
Man, and his stillness stiller
Than song of an unborn bird.
Night chilled, and grew chiller.
The tent was not need for him;
Nor destiny. Not a star
Walked, a smooth-limbed god and slim,
But things were what they are.
Less than stillness, silence
He winged beyond Creation, he
Full with indifference,
And head bent near to his knee.
Lips murmured a thought's births;
And curled, and eased. The oldness
Of sun and man he knew, earth's
Coldness and goldenness.
Night aged and pearled
To a dawn; yet the philosopher
Turned the globed world
Colder and goldener.

FOUR LAKES' DAYS

I

Summery Windermere, sweet lake!
Where eyes wake wide to see again
And the fields, O clover earth-breath take
On the roads, in the warmed-away rain;
Then Far, and then Near Sawrey
Are; foot-increasing Esthwaite
Water is, O the green! far, see
What you will, walk. There's a late
Spell on the hills, will break not,
A gentlest mystery like a fate,
Something serene, you cannot take not
On Hawkshead Hills. Blare on the air,
Bagpipes! 'tis the merry-hearted note—
Thus you change, hill and valley share,
Feel each, know none by rote.

Young ferns by time-retaining tarn
Lose something of goldenness: height
Down-dooms, moves along the tarn;
(In poised fragility, my fright)
My heart leapt down, so with gloom
Leaded and time-encased, I shied
From the tremor, the small room
Of (bleak tarn!) all that has died.
Glen Mary unsilences; slight
Whispery waterfall, then such
Tossing laughter as can fight
Rocks, and overcome them much.
On heart-beat and feet go, he, I,
Who than happiness happier be
Keeping earth joined to the sky,
Locking, in love, earth and the sea.

But, topped, stop. Somehow
Miser it, lest the mind

Race beyond the senses' pace
And the whole being nothing find.
Valley-volleying blackbird (how new)
Another flier has; sheep can
With a strange passion bleat; there grew
In spider's earth-sight something of man.

In air-shiver against white wall
A flower, blue, incandescent
Does dance on the eye-ball;
It makes beauty effervescent.
At Fell Foot (charm) farm. Rob
This, clover-over-the-dales;
Smells of cows; good hob–
Nail, spike my ear! and kissing pails.

Till the twilight folds and all's
As blue as the bluewashed walls.

II

Point-pushed by the wind, rain
With penetrating excellency
Scrawls all the hills, writes plain
Its unread-yet charactery.
Now in wet morning be, begin,
Wrynose Pass to Cockley Beck,
Ramp, suck the thick damps in,
From cheek no lightest rain-flake fleck,
But all is watery world-of-love.
Up (darkly, slowly) Hard Knott Pass
To the bleak. There, mind's of
Not the tenderness of the grass.

So lean the weather-fabricked eye
And build the chilly-mossy dell
From universal motleyed sky
To this ten feet of world, of fell.

8

Not the woundedness of the soul;
Of desire, the joy. And, soaked all day,
Sky, earth make each other whole,
You between, cold mote in the gray.
Leave Duddon for Woolpack Inn.
'The way's by the little stone men
Over the moor.' She said, whose thin
Skin at the cheeks thought youth again;
'Turn left by yon leafy tree.'
Her winter she turned to the valley.
So he, I go, we
Slow to the pelt-rain-drum's rally-
Of-loneliness in rained-on weather.
Fluffed, but damp, and moving, boulders
Bleat, all hill-huddled together.
Hump up their pointed shoulders
The little men, fearful, who dare not
The Druid haunt; but they can lead
By bitter Burnmoor tarn! and care not
For anything but the earth-need.

The low skies lip the lost lake,
I quiver to taste the dead sight,
But to my lips the colour, livid, take—
Sca Fell scales his hoar height!
And I make to the bleak Screes
Where the rain is alive, though dark,
Thread through wild rain on my knees,
To Wast Water. The misty mark

Calls. And I sleep to hear
The fine, moist lungs of atmosphere.

III

Labour them, these drenched heights.
Put foot to dusky sky-dropped pearls

That rope the grass with their frail lights,
Spring from the shelly strength, it hurls
You up! and at Windy Gap
A born cloud will take you, lake you
Around in its fresh frothy sap;
And the pulsation will shake you.
Speedy steady wind-of-mist
Blew up the side, the ledge combed it;
Stream out the fledgling and me-kissed
Spumy streamers and I roamed it,
Happy-shawled, the high valley air
In the condensing and tight cloak,
My new body-skin and fay's hair;
And clean, and wet, and warm soak.
See it cling its youth together
Slowly over the valley rolling,
Fluffing a myriad down-feather,
Coyly downward. Whole-bring
Your ghosts here; and all-take
The gleams here, sense, sense,
Through the sun's eye of this lake;
Till Black Sail's sound of waterfall
Tries; and you are down the hill.

The Northern man, with a ripe lung,
Found vagabond, can fill
A dale with his nutty tongue.

Climb up Scarth Gap, lit
Dismally by foreboding clouds,
And at the top, the desolate pit
Portends with dull baleful shrouds;
Must wear the rock, bound solitude
Like a leaden, and forlorn hope,
Or like a worn, lost brood
Of these, that crowd and grope,
Stumble, fumble, mell down

Into the sun-on-the-hill sheer
Prospect sweet and, there, town,
The final soft of Buttermere.

Lie on stone bridge and be lulled
A long time in the gold shocks
Of trembling stream; till all's dulled
But the sombre shine of under rocks.

That light will deepen and well, in
Snug sleep, under Helvellyn.

IV

Contend with steepness, the dark air
Unshapes the mountain as it grows
More misty-moody. Pace it bare
Up into the stormy throes,
A certain waft of rotting fern
Gone by in the severe, clean
Feel of wild electric burn,
Needles of ice-like rain in the sheen
Of darkening lambency and surrounding
Tempest, as the clashing sound-of-
Clamour clouds, mountain-bounding
Grow to a groan through the pound of
Immense winds in the gloom-nude space.
Lost, and aware, with hard, clear
Climb, steer at the force in your face
Through the tightening, soft, obscure
Orbit of dark inhospitality,
And the powerful blood dominant leaps
Up at the dreadful world-majesty,
At the bleak raging energy it keeps
Stupendous tempo striking the rock
With slant steel feet, with mailed flesh
Commands the creep-on fog, block, block

With blackest waste its trembling, fresh,
Impetuous urge. At the felt top
The violent wind and rain shake
Exultation, and time ceases,
The pulses jangle as if to break,
The churning chaos choas increases,
A surge of demonic energy unites
Blood and the bitter world-vitality
As the flaying and flayed being ignites
In elemental passion intensity
Satanic, angelic, one harmony
Of immense glory like fire clinging
A blaze of terrible immediacy
The wild blood of freedom singing
Descending the clanging cliff rungs
By its own impulse hurled,
Shouting with soundless, taut lungs
Into the powerful, dark world,
Catharsis of the soul triumphant
Through suffering and effort, death
Purged from the bones militant,
The gift of God, pure mortal breath.

There's tremble in all solitude,
Be at, in this: it's in you, times.
Nothing of evil or of good,
But the quiverness, sometimes.
It is given. Yet, feel-less
How often, we, fallen fell adown
From our free skyey realness,
Too close to men-mesh and town.

Mystery is for ever the same.
Wake into the life of flame.

ODE TO SILENCE

Be Silence from your dismal cave abroad,
Nor moist rocks sweet with rain, nor mosses
Lately drenched, that swell the broken sky,
Postpone, now mellow doors of fragrant air,
Your stealthy moving cloak here where we die.
Come from your grot, the last power left to laud.
For she lies dead, and agony can no more.
Sun light on ebon hair. Be one witness,
Though strange noon, with hurtful whisper-shine
Crowds down the sky, on golden chords that tear.
With surety firm and deep as ocean's roar,
From dauntless throat, that moans and wails all year,
Be hourly spreading tenderness no tongue
Can call, O steadfast force in sojourn here,
From region dim of love, but something young.
So shall we join, so both pour out our store,
By tarns of melancholy heights of hills,
Or on torn cringeing sands of deserts, there,
Or barren knolls, abrupt from wide valleys clear,
Or wandering among the tribes of snow.
How have I kept the eagle eye for you!
Discoverer ever haunted with new worlds.
Consent to me, and dare to seem
The fervent armour of my skin.
Thus quell the risen anguish rude.
Stillness, be my plight: mate with song.
Then action's fortunes must, as avid wires
Reverberate, be vocal calm.
When the bumble-bee drew down the clover nod,
But gentle, shook the dozing afternoon
With his dry burn, and all the royal compeers
Helped moan the cooling earth to your full boon,
O Shadow born! then was an end of tears.
Orchard's grass, and the Septembered sod
Slumbered, and covered over were green fears.

No foam stirs forth from out the milkweed pod.
No bird can step, but earth's whole nature hears
Its little spring, and pleasured, seems to croon.
O joyless land, of wilderness
Increasing strong, dread solitude,
And now the sky brazen and stiff.
Why must all ways be closed
Against me miserable
From the fount of joy
And springs of innocence,
I left in this harsh world alive.
We die with her as all close with their dead.
Now walks full midnight, O make no moan for her,
Now bind time's mantle, the dark cloth, in folds
Around, to spirit same as fleshly stir;
That the mutual control a hoped life hold
In shroud, the flowered interment of the years.
Whence comes some nameless peace that still enfolds,
Though the world sadden, and new pain has bled
Ere the old is stilled. With Silence, in dank wolds,
Or humid meres, mostly worship her.
Ah, wounded more, in earth's increasing ways,
Be sound the kind counterpart of need.
Hear the grieving Chinese gongs emball
Woe's Aeon, and roll it in a hollow note
On the world. And now the softer drone
Of the love-lute in Sumatra, yet remote.
There, against Girgenti's temples thrown,
Some careless boys for Pan pipe simple lays,
And sing Sicilian airs, too fleetly flown.
Slow flows the moist Alpine horn, and falls
Mournful from the mountain's lucid stone,
The perfect lament, and threne of sorrow's throat.
All sound is ladened! when human means
Reach forth. Then listen the earth sings
Faithful to itself, unknown.
Clairvoyant, be as a cool sea shell,

14

Constrained, with hurtless ear
Whirled on the gentle roar,
To the impersonal impersonal
That delicate ravishments in the mind come.
Yet plainly I hear our unyielding voices cry
In the long suffering that knows no close,
Nor earth smothers, to rise on lark note through,
Or like the crickets sing. Aye, even our woes
Recumbent, seem to fever forth with clue
Of other sorrw, and our instruments sigh
To wean us from that deeper natal rue.
Soft mineral, instruct the bell to die;
Fibres, be wood of waxen-sculpted stop;
Dry thong be lyric, in catgut's chilled throes.
Earth sees her no longer, no more her beauty
Lingers. Love lights us no more. No longer
Love overcomes us, and maims us, and will kill.
Be earth of strength to make us the stronger.
Be this the will, we selfed, enriched in her will.
To move, through this midnight, somewhat free,
Under the sumptuous warping cloak, until,
As sap can yearn though Destiny stand a tree,
We wander, but will kiss the white star light,
O sovereign lovely Silence accept, be me.
Your kingdom's in the wayward wind
In waft of ease above March bulbs,
Where too you pulse.
The sea-mew finds you after his call.
The dusk ferns
Are yours, when the sun's become their own;
And a green shine they lend to the small brook;
Then thronging minnows, like lights, are seen to play.
O joyful land, of solitude
Brimming bright, loved rarity,
And the heart outgrows passion and pain.
Vibrance is on the violet now.
See, the whole sky trembles, mirroring

One simple flower. Stillness quivers, and forces
Vision inward from the eye.
There is one way, to creep up the purple tunnel
And draw on that tender canopy;
Thence from happy shadowy
Contentment, look outward
Secure, with world bright immaculate eye.
Quivery are in all earth and ocean Silences,
Whether fire's or ice's, or loams' or stones'.
By smallest ellwand touch the deep moonglade
Until our harmless obsequies are done.
Firstly luminous in the beginning,
All things sing Silence to the perfect sun.
Flowing starlight, like a frozen thing,
And mountain, like a fusion not yet made,
The rayless mind relume, and in it ring
(By gyres on gyres of harmonies quickening)
In changeless radiancy that casts no shade:
Universal fire mothers its piping light.

THE RETURN OF ODYSSEUS

My disappointments, large as capsized tugs,
Pull no more the big ship of love.
My eye of fire, motionable coal,
That saw such salutations of the wharf
(And piped a hoot to the woman's tooting colours)
Is out, and weeps annual delugings;
My engines bucked and wrung. My
Timbers' splinters plaything water patterns,
My sturdy and few tons have gone to roost
Under the infant waters of the sea,
That never grew to man's curious abyss,
Nor walked a brain in the aching shape of dream
Along the narrow deck of singing timber:
What a boisterous business, to be a guide.
But in the nest of nature's slippery scope,

(Where of God the pressure is so tight)
Elongate eyes like spears would be
Would spasm the cranky juice of hatred's darlings,
Spend the volatile dyes of woman's dress,
Precursor of the sorrow's end. Unheard
The cock in the waters crows, Good Night,
There is nothing to do but be swayed by the sea.

While with gynæcologistic hands
My hempen fingers purged, I had eased
To birth my iron monstrous love.

'WHERE ARE THOSE HIGH AND HAUNTING SKIES'

Where are those high and haunting skies,
Higher than the see-through wind? Where are
The rocky springs beyond desire? And where
The sudden source of purity?

Now they are gone again. Though world
Decrease the wraith-like eye so holy,
And bring a summer in, and with it folly,
Though the senses bless and quell,

I would not with such blessings be beguiled.
But seek an image far more dear. Oh where
Has gone that madness wild? Where stays
The abrupt essence and the final shield?

SUITE IN PRISON

I

The grave's a seed will get some monstrous bloom
To drowse the mind with warm eternity
And nourish the stupor on unending time
Until dull numbness gluts the feeling power,
Then close its final grandeur down and dark
The last clear beams of bright mortality.

The skeletons of lovers, let them then
Rejoice since no hot hurried consummation
Vagues their joy or surfeits fond desire;
The earth is long at marrying the bone.
Let fleshless jaws make smile at butterflies
And humour have and jest at humming birds
Since these have got their moment wooing done
Before the axle of the earth has turned,
Or harvested a wormy crop of moons;
But they sink deeper in abysmal love
And in the slowest satisfying lust
Are tightened, till they get great regiments
Of babes with petal spears to pierce the air.

Bull worms will probe my sockets, move among
The loosening channels of my brain for food,
But how, when they recoil from some hard tissue
Bitter to their taste, shall they surmise
Stout Plato with his knotted dream of God,
How guess the Christ-idea marbled there?
And when the brain's last particle is worm
How shall the worms high triumph celebrate
And how rejoice for hugy victory?

II

That which is at your East, your heart, I cannot
Tell how it shapes with mine, thus to admit
By a fine dexterity our equal pleasures.
It is too bad we do not in Universals sit.
You have your South, I mine, joyful, painful,
To which each goes; and gets from, fangs' stings.
But we must burn there, that some fire stay
Which nested the thick Phoenix-matrix of things.
It seems the West is to us both plainest;
Escapeless; stopless and blackest. We are so
Blights in the tall scent of this crouching hunger.
To devour the race. Thus let another grow.
Hearts, testicles and gall, we defeat
At the head, our Winter; with eyes' changeless white
Got gazing in that mirror, supernal earth.
And there, whirl, chilly, swift, and replete.
Yet in all this intercommunion I feel with you,
I weep the foreign waste of your close brow.
Leave me. Alone I can pluck up a looking-glass,
And say my own eyes to my own eyes are, now.

III

I turned where the sun came up and he held my hand,
The fair lily of lingering in his hand.
Where he walked, freshening over the land
He plucked me up from the fragrant land.
His breast was smooth and smelt like pleasure,
Joyfully I bent and kissed my treasure.
His skin rippled senstive as fingers;
Peacefully I touched together my fingers.
I lay on the green earth moist as grass
And I reached, but the sky was not grass,
O weep alone! I wept on the world,

And weeping came back over the waste world.
I turned face down to the mean loam,
The worms all choired come home.
Than a flame lit in my thighs and feet;
I ran away, yet not feeling my feet.

I turned where the sun went down the dark,
And a hand was on me, hairy and dark!
As he rushed, looming out from the earth
He tore me up from the fervent earth.
And a monstrous wailing destroyed the air.
But no one was there; no one in the air.
Screaming, faint, I peered down to the ground
Praying for the womb of the black ground.

IV

Incarcerating sunlight, world damnation,
Now am I crisped in the great blight of all.
Incantation of might, my personal laudation,
To you, now, at last, from me must befall.
Power blest, against blasphemy Hate
Lies down a spent child with not one sigh.
Beyond hope, we are destroyed to create
Anew, paeons in the killing sky.
I cannot combust off the skin-grown sun.
Darkness proffers no help and cold no chill.
Pitiful in incandescence, Furies of action
Dancing, their impotent conquests distil.
Angels of solitude, immured in monuments
Of cruel hours, malignant despairs instil.
Vision without memory has no portents.
Future, terrible and evil, be me, I will.

V

THE BATH ROOM

Shoes, table, light, and telephone.
The polished hides point to the wall.
They want their grass. And some antique cow
Likes the mild wash of the pagoda bells.
And I'll on no pillow moan,
Nor in the balefires fight,
But call up the Wolf, and hear him rasp,
And say I'm in this room tonight.
Some other storm, come cut the wires down!

Those kissing hides. This glass-refracted light;
Whose incestuous brother fails his love to tell;
And if it were possible, I'd swing on a willow.

Khubla, but with mailed boots. Akhnaton
With a taper. The Beast ravens, without our change.
I stand bolt upright, reading the paper.

The table wears its faultless, white surplice.
Fixes, waiting for an elbow.
Shall I dart down my diamond eye
And all its fibres try to tear?

No, I shall stand in the middle of this room
And not note the soles of my feet.
Not front the swords of the light.
Not brave the yells of the Wolf.
Not be adamant-minded in the middle
Of this doom. But with sweet acclaim
(By the bedside) modulate the name
(Light, shoes, and telephone),
Of the albino crow, of the table.

VI

Lord, stabilise me. My legs
Fail in the white crevasses. My hands
To thin husks are twisted. My bones
In the high winds can stand no more.

Blood, that built the heavy world,
Curls in the shells it built,
Thins and congeals. Its cold heart
Would turn to the earth, its home.

Eyes, the last levellers of the world,
Have lost direction and show false,
The rare air them avails
With many ways of ice and snow.

Of glassy temples crinkling white
Glittering all the icy air,
As cold a city and as bright
As flicks in the tight air of a child.

Lord, stabilise me. My pride
Walks with inebriate legs. My desire
Eats up the whole world. My love
In its big excess destroys me.

Lord, qualify me. To see
Pride floored like a marionette;
Desire fulfilled like a marriage;
And love of sweet use in the world.

Lord, admit me. To rejoice
Even within the abysm of suffering.
Contain me in Thine atmosphere, Lord.
And in the dancing of Thine ecstasy.

THE GROUNDHOG

In June, amid the golden fields,
I saw a groundhog lying dead.
Dead lay he; my senses shook,
And mind outshot our naked frailty.
There lowly in the vigorous summer
His form began its senseless change,
And made my senses waver dim
Seeing nature ferocious in him.
Inspecting close his maggots' might
And seething cauldron of his being,
Half with loathing, half with a strange love,
I poked him with an angry stick.
The fever arose, became a flame
And Vigour circumscribed the skies,
Immense energy in the sun,
And through my frame a sunless trembling.
My stick had done nor good nor harm.
Then stood I silent in the day
Watching the object, as before;
And kept my reverence for knowledge
Trying for control, to be still,
To quell the passion of the blood;
Until I had bent down on my knees
Praying for joy in the sight of decay.
And so I left; and I returned
In Autumn strict of eye, to see
The sap gone out of the groundhog,
But the bony sodden hulk remained.
But the year had lost its meaning,
And in intellectual chains
I lost both love and loathing,
Mured up in the wall of wisdom.
Another summer took the fields again
Massive and burning, full of life,
But when I chanced upon the spot

There was only a little hair left,
And bones bleaching in the sunlight
Beautiful as architecture;
I watched them like a geometer,
And cut a walking stick from a birch.
It has been three years, now.
There is no sign of the groundhog.
I stood there in the whirling summer,
My hand capped a withered heart,
And thought of China and of Greece,
Of Alexander in his tent;
Of Montaigne in his tower,
Of Saint Theresa in her wild lament.

THE RAPE OF THE CATARACT

King Prajadhipok his cataract
Doctor Samson held within his hand
Walking down an aisle of attendant
Princes, and maids, awaiting the outcome.

There, of obscured Oriental vision
Was an end, and caul; breathlessly
The ten Private Secretaries, the General,
The Admiral, the established flunkies

Saw the veil drawn off the ancient East,
Rejoiced at Oriental Americanisation,
Made delicate bows, appropriate sighs;
While below the Court clicked billiard balls.

The Royal surgery (decor by Le Corbusier),
As palace made into an hospital
Reeked not at all of the lotus blossom,
Nor of perfumed cigars, but stank

Very slightly of an overt efficiency,
The time element increasing anxiety
To see, or not to see, if to be seen
(From Bangkok to New York) the western world

While the dapper Minister in excelsis
Haunting each crevice of the vast palace
With feline graces, a misplaced Machiavelli,
Sipped his calculated Eastern tea

Having a score or more of gentle daggers
Concealed in his innate gentility
But not a fleck or flash of any eyelash
Took aim and stock of failure, of success;

Djin comes in. His Excellency whines.
And then my suave and gentle Chirasakti,
Adopting a smile, is waited upon
By Smaksman like a bird with bugle eyes.

Now hail the light! His Majesty appears
As the neat procession of success begins
Flanked by attendants in a soft fatigue
As the King's hayfever shakes a silken handkerchief.

The President of the United States must wait,
And while his jaws contract, His Majesty
Upon the lush parterre with Djin and Ajja
Makes timid, deft essays with a model airplane

Its little rubber bands expanding
As with a gentle whirr it takes the air
To soar ten feet, then, glory-spent, it dashes
At the feet of the King's frightened Secretary.

And now the Royal Party moves in state
To the mansion's chief balcony; there arrayed

It views the proper burst and display of fireworks'
Perfunctive poppings on the feckless earth below

For Western vision now succeeds the Eastern,
Radios, picture palaces, and television
Make a regal triumph at the Royal Court,
And Their Majesties retire, amused and pleased

To see the President, and shake his hand
For fresh experiences in the West all eager
And then departing in a bolt of motors
They seek the City, while still Buddha smiles.

1934

Caught upon a thousand thorns, I sing,
Like a rag in the wind,
Caught in the blares of the automobile horns
And on the falling airplane's wing.
Caught napping in my study
Among a thousand books of poetry.

Doing the same thing over and over again
Brings about an obliteration of pain.
Each day dies in a paper litter
As the heart becomes less like a rapier.
In complexity, feeling myself absurd
Dictating an arbitrary word,

My self my own worst enemy,
Hunting the past through all its fears,
That on the brain that glory burst
Bombing a ragged future's story,
Caught in iron individuality
As in the backwash of a sea

Knowing not whether to fight out,
Or keep silent; to talk about the weather,

Or rage again through wrong and right,
Knowing knowledge is a norm of nothing,
And I have been to the Eastern seas
And walked on all the Hebrides.

Ashamed of loving a long-practised selfhood,
Lost in a luxury of speculation,
At the straight grain of a pipe I stare
And spit upon all worlds of Spain;
Time like a certain sedative
Quelling the growth of the purpose tree.

Aware of the futility of action,
Of the futility of prayer aware,
Trying to pry from the vest of poetry
The golden heart of mankind's deep despair,
Unworthy of a simple love
In august, elected worlds to move

Stern, pliant in the modern world, I sing,
Afraid of nothing and afraid of everything,
Curtailing joy, withholding irony,
Pleased to condemn contemporaneity
Seeking the reality, skirting
The dangerous absolutes of fear and hope,

And I have eased reality and fiction
Into a kind of intellectual fruition
Strength in solitude, life in death,
Compassion by suffering, love in strife,
And ever and still the weight of mystery
Arrows a way between my words and me.

In a hard intellectual light
I will kill all delight,
And I will build a citadel
Too beautiful to tell

O too austere to tell
And far too beautiful to see,
Whose evident distance
I will call the best of me.

And this light of intellect
Will shine on all my desires,
It will my flesh protect
And flare my bold constant fires,

For the hard intellectual light
Will lay the flesh with nails.
And it will keep the world bright
And closed the body's soft jails.

And from this fair edifice
I shall see, as my eyes blaze,
The moral grandeur of man
Animating all his days.

And peace will marry purpose,
And purity married to grace
Will make the human absolute
As sweet as the human face.

Until my hard vision blears,
And Poverty and Death return
In organ music like the years,
Making the spirit leap, and burn

For the hard intellectual light
That kills all delight
And brings the solemn, inward pain
Of truth into the heart again.

'MY BONES FLEW APART'

My bones flew apart. They flew to the sky,
And as great knockers knocked the air.
The bone God said, Come in,
My soul flew in small and dry.

It is the vale of lamentation,
He said, where Peace never comes,
I thought it was the end for ever
And the forgiving song of no tone.

I was big with thick desire,
I had love like leopard's love,
Time was, time is, time will be,
But see me change I cried horrified

And these dead parts, as dead birds
Tried now wings new and tireless
Winging the fine strife of life
And sudden strength all blind and wry.

We are in the kingdom of awe
My bones cried, my bones sanctified
That blood and bitten mind ne'er dealt
And a cold mist arose around

And the God sighing: I am fiction
I am Intellectual Fruition
Forever a folder open of portals,
The eerie trial of infinite tears

Till my spirit, gasping and sunk
Bereaved of all complexity
When the sun came surging, shook
Back into its knitted unison

Then truth is but simplicity
I cried, revivified, sanctified
And turning to my neighbour alive
Said, I will give my life to you.

THE TRANSFER

Jumping from the certain earth, the river so
Exciting me with the speed of ice and thaw,
From all that solid is, to that which flows,
And spring was in my blood lusty witchery,
It floated before me, island of all excellence,
Crystalline in purity and rarity,
Carried in a slow and bold magnificence
Along the coursing and the swirling river
Like muscles flexing in the backs bent
Of million members of humanity,
A carriage huge, and in that a casket,
Lapped with wavelets tinkling, thence tossed back
By bright and diamond hardness of the gem;
In which I saw the visionary gleam.
And saw my desires fixed in that erected ice
Like giants standing still with mighty eyes,
So alive they had achieved full solitude,
Forms immovable, calm and unassailable,
Whose ruddy breasts breathed the river's rhythm
And their flesh flashed a radiating energy
Shimmering the finely humming air of April,
Then I stood dazzled in an ecstasy
Knowing waste and loss were overborne.
And now my blood so fiercely burned with desire

All was in me lightness elevation and love
While the air's silk beams could bruise
Fingers so keen, though like a steel blade
They would cut the fibrous element between,
Feeling the fine white edges of the ice,
Wherein embodied power and wisdom stood.
Action suspended above both evil and good.

And the ladened river its impending gift
Flowed into my eyes nearer and yet nearer,
Bore onward toward the bank on which I stood;
And when it loomed before me like reality,
I leaped, and a clashing thrashing of the waters
Sent the white ship toward the middle deep
As I crouched in humility, shaking with dread
My being feeling a strangeness like electricity,
The exultation of a world outgrown,
As when a sick man comes to health again.
And as the clear spirit, riding on existence pure,
Lifts its eyes unto the unfathomable blue.
I slipped inside the ice, avid, and put on
Transparency like a flawless feeling of myself
And I saw the world no more as it is
But I had become the vision of my desires,
The island of ice, the ship, the casket, and gem,
Floating on the teeming river's force,
Standing statuesque in the shape of man,
Heroic in the virtue and the power that I adored.
Hour, hour, hour I so flowed.
Time has no running, and it runs not away;
Realisation of being is unchanging felicity;
Desire fulfilled, the end of human action,
Releases in the deep dream of contemplation
The natal awareness of eternal harmony,
And I had longed to leap from solid land
Upon the river that bore ideality,
And felt my brotherhood with heroic men.

Serenely down wide water through the land,
The cake of ice its silent progress making,
See me in the flawless veils of ice
That strange monument, that lifted shape,
That powerful tower in iridescence gleaming,
And though the river moves, I feel no pulse.
She rode in triumph. But the light pierced in,
Shimmering to a flag of gliding colours
And these shivering chimed frostily together
Like swords clashing, kindling spells of music;
Creating pressures, a stealthy sense of dread.
The sun leaned down upon the middle sky,
Then I felt a cleaving, a shiver, a shock,
And like the crumbling of a sheet of cliff
A slab of sliding ice split off and sank.
I looked outward startled, feeling my body's pain,
The opening of a wound that cannot heal,
Like loss of a limb to one who feels it still,
Another rhythm, a breast of the ice tore loose,
Canting into the waves with slipping ease,
The river gathered, the sun pressed closer down
The river began to churn with struggling waves
A stroke like lightning hit and stripped from the body
The last thin scales and brittle veils of ice
Then I, in this cold pitfall snared,
With nervy grip, grasping the raft of ice,
Saw the monstrous passage of the world,
Naked I walked the small confines of hope.
Panic was in me, sickness and clawing death
And like a feverish animal in a cage
I walked forward and then backward frantic,
Another piece of the ice cracked off, another,
Until I had bare room to stand upon.
The land no hope for now; cruel land, she loomed
Like loss, like ruction, like a great prize,
And must for this extreme mistake I pay
(My feet ache, my hands sting, my blood

Congeals in fear immense and caged hatred)
With my whole life, drowned in the swirling tide?
There was a little quicker pace in the river,
A dissonance of the unequal stress and strain
Of eddies in the flood, I turned backward
And I saw the force no man can stop
The river looming sullen and uncontrollable
And coming with an ever faster motion
And clashing, hissing, and with massive turmoil
It bore in a vicious surging on again
And I turned forward again livid in panic
I saw the whole force of the mighty surge
Caught in a gigantic breach of clashing,
A whirlpool churning tons of cracking ice
And as the ice I stood on gathered impetus
The baleful scene froze on my mind, I saw
Ice cakes like shapes of a ghastly ocean
Like living men reach up in air
Some with cumbersome motion poised in air
Before they slid down under the tortured mass
While others cracked to bits on the seething surface
Dashed in fragments over the frozen dam
To the slithering downward chaos, abyss delirious
I heard the gashing, slashing of fierce souls,
The quaking shrieking chorus of destruction,
And the eerie sound beyond the world.

Threre was a tree that hung from out the bank
Austere, as if it were a living man.
An arm he held over the flood, and as my craft
Leaped, shivering into death, I leaped to life.

REQUEST FOR OFFERING

Loose the baleful lion, snap
The frosty bars down from his cage
And unclasp the virgin pap
Of the white world to his rage.

See the innocent breast deny
But the bellowing shake down the air
Shudders of passion out of the sky
To shock, mangle and maim, tear.

Under the actual talons see
Virginal white and the black paw
Poised to slash on mystery
The five hates of a claw.

Amaze your eyes now, hard
Is the marble pap of the world
And the baleful lion regard
With the claws of the paw curled.

Loose the baleful lion, snap
The frosty bars down from his cage
And unclasp the virgin pap
Of the white world to his rage.

NECESSITY

See her austere beauty bend;
And her desperate eyes are still.
She walks in terror. And the deep fear
Fascinates me under her will.
The full power of nature unfolds,
Using her like a blind seed.
Now in the awful stillness, watch,
She walks with a panther's need,

Destruction, immediate, violent!
But her lips are livid and chill.
When will the hurt mouth quiver,
The great agony break, and lie still?

Look into her sober eyes, they
Coil sorrow; but the grown malice
Will not spend. There is no hope
To-morrow any hope can be.
It is too late to will; and time
Will blunder its bold meaning
Into her blood, deeper, and more deep,
Then plunder. Be away;
You will not dare her then,
To see the coils of hate uncoil
And the wrought bitterness snap
Free. Go into the woods
And praise the inhuman sun;
Unless wells strangely in you
Pity. The woman will make
The blind cells not to grow.

There is not anything to be sure of,
If destiny will destroy or will build;
And there is nothing to be learned of love
That will not suffer change, or be killed.
We are always about to be used
And are used by nature, without escape,
Save that our wills are with hers fused
And we would impregnate her with our shape.
But in the great moments of being, something
Beyond our wills, is the prime mover
And we do not deny this when we bring
Passionate love to a woman, as a lover;
Since we are compelled by a hid purpose
We cannot control, if joyful or morose.

It is a terrible thrall to be alone,
With all joy there, and destroying fate
Slicing the flesh, hot fangs on the bone;
The intense quality of desire
Blasphemes, and it is at fault to the core.
Silence in bitterness is the hardest thing;
But nobler to ask the fire to burn more,
If the mind can endure, and can sing.
Even beyond joy and despair are spun
Unutterable remoteness in the air,
Intolerable nearness in the sun,
And the separateness of each man in his lair.

THE SCARF OF JUNE

I do not feel spring's windy ingress
In the thickets of my heart
For yet still winter with breast-bands
Of ice, imprisons me.
I slumber through time's weariness,
Apart from phantoms, far apart
From sun's young running down warm lands
With flame for flower and tree.

But when my roots must cringe and thrill
Feeling sharp heat dig down in earth,
My heart will only, leaping, kiss
June's woven scarf that strangled me.
Spirit that animates man and hill
Will work in me hot seasonal birth,
But I, insensible to this,
Will like a cold stone be.

Thus time can do no harm, so let
Him come, through roots and branches creeping,
Bringing spring and winter in
And hurrying life away.

For I am where no sun shone yet,
And am earth's inner being keeping,
Locked in this lone discipline
Against the world's decay.

EXPERIENCE EVOKED

Now come to me all men
With savagery and innocence,
With axe to chop the fir tree,
Or seed, small, for the immense
Sewing of earth with old Rose.
Now come all men, arrayed
With the colours of the garden
Around them where they stayed
Till bone began to harden
Under the thinning of the nose.
Come all men, unto whom
Wind was a snarling wire whip
In the contusions of a doom
And with red flecks on their lip
They leaped up, danced, grew tall.
Come all, the babe bound
In terror and panic cry;
Or an old man found
With a skylark in his eye.
Come, harsh shroud over all.

TWO LOVES

That her serene influence should spread
An afternoon of soft autumnal light
Is to my heart not unaccountable
For she was young, and is not dead.
And still her cheek is red and white.

But that this stealthy still insistent power
Pervades my mind and will not slumber me
Is delicate woe and glory hard to bear;
Her life lives in a ghost-wrought hour,
From whose chill spirit I am not free.

The one was willow to an ardent touch
And she was mood that had a right to die.
But she, the other, the passion of my mind
Long-living still, does overmuch
Come from the dead, and from the sky.

BURDEN

Whoever lives beside a mountain knows,
Although he dares not speak it out, that he
Must always carry on his heart the snows
That burden down the trees. And never the sea
Will rush around him cool, like snow-cool air,
And carry him and lift him like a leaf.
He will not find this lightness anywhere
Since mountains brood, they hold dark league with grief.

The pine trees never tire of moving down
The slopes to meet him, pointing up from town
Beyond the tree-line to the rigid peaks.
The mountain holds him though it never speaks.
He scrambles over boulders on his knees
Trying to reach the summit, like the trees.

'IN PRISONS OF ESTABLISHED CRAZE'

In prisons of established craze
Hear the sane tread without noise
Whose songs no iron walls will raze
Though hearts are as of girls or boys.

By the waters burning clear
Where sheds of men are only seen,
Accept eloquent time, and revere
The silence of the great machine.
On the sweet earth green and moist
When vainglorious cities magnify,
The senseless dissonance will foist
As witless on the shining sky.
There is some stealth in rhythm yet
Albeit an even breath is not.
In the mind is a gauge set,
Lest the blood spill, and blot.

THE LARGESS

With cicada's nymphal skin
So have I meetings made,
Let down my eyes to him,
With fear upon that thin shade.

Lest the look I gave
Was death's loving me,
To every memory have,
That himself he see.

Yet O marvellous crispness,
Dun, but perfect structure,
Thin as matter is,
It has its wondrous lure.

And took it in my grassy feel,
That cold, that final form,
If still it be the same;
Alert to a hoped harm.

Where have you gone, slight being
Whose brown monument

Mirror makes of wings
Yet in a damp tenement.

Can I among winds lose you
When vibrant is all air?
Must I not use you
Then in every desire?

Do treble drums a changing
Ecstasy keep fresh;
Insistent, sing to me,
Over fields of August.

It has not denied my mind,
But no sign has made,
Bleak, delicate, defined
And crinkled husk once life had.

My eyes soothe over him,
My hand trembles with force.
What eternal hovers in
Him: speak, are you corpse?

'WHEN DORIS DANCED'

When Doris danced under the oak tree
The sun himself might wish to see,
Might bend beneath those lovers, leaves,
While her her virgin step she weaves
And envious cast his famous hue
To make her daft, yet win her too.

When Doris danced under the oak tree
Slow John, so stormed in heart, at sea
Gone all his store, a wreck he lay.
But on the ground the sun-beams play.
They lit his face in such degree
Doris lay down, all out of pity.

'THE CRITIC WITH HIS PAINED EYE'

The critic with his pained eye
Cannot my source espy
For truly and purely to eye it
He would have as Critic to die.

I with joyful vision see,
I cannot his purpose acquire.
For if the Critic were truly free
He would love, and not be a liar.

THE YOUNG HUNTER

Here gunned he homeward in the birdy breast;
Muscles full of baskets spread a picnic.
Let the palping trickle, the wing's blood's best
Be, as it was, his hunting heart's quick.

And I allude a mary-gold's guts
Are no match, brass heart, for yours.
Nor dug themselves in magnifying ruts
Where passes to repast, hulk that stirs.

Observe the revel of the nerveless finger.
Praise the young; lo, praise the dead.
It changed none the mysterious transfer
That long since knelled a questing in the head.

'WHEN GOLDEN FLIES UPON MY CARCASS COME'

When golden flies upon my carcass come,
Those pretty monsters, shining globules
Like tautened oily suns, and congregate
Fixing their several gems upon one core
That shines a blossom then of burning gold,

'Tis as the sun's burning glass and diadem
They work, at the first chance of rotten flesh,
And, senseless little messengers of time,
Some beauty keep even at the guts of things,
Which is a fox caught, and I watch the flies.

'NOW IS THE AIR MADE OF CHIMING BALLS'

Now is the air made of chiming balls.
The stormcloud, wizened, has rolled its rind away.
Now is the eye with hill and valley laved
And the seeds, assuaged, peep from the nested spray.
The bluebird drops from a bough. The speckled meadow-lark
Springs in his lithe array. Fresh air
Blesses the vanished tear; the bunched anguish.
The laughing balls their joyful pleasure tear.
Renewed is the whole world and the sun
Begins to dress with warmth again every thing.
The lettuce in pale burn; the burdock tightening;
And naked necks of craning fledglings.

THE CHILD

Five years of cringing child small,
Bewildered by the barren wall.
Laughed Atlantic its heated summer,
This I do not remember.
O the fearful cliff wall,
And the fearful sapling nigh.
And the dove that will hurt me,
And the great hand of the sky.

'LET THE TIGHT LIZARD ON THE WALL'

Let the tight lizard on the wall
No green spike see: but tickled eyes
Sparkle up at the sun's dazzle,
Lungs, lithe, strike no sigh.

So to be instant adamant,
Pipped, a nimbus exhales,
Of exquisite leaf-lave,
And he contains time's fail.

And he durst himself absorb,
The canker pelts to defy,
Blot of dark he to exude,
That marriage of the sky.

Quiver no joy to me, I see
The world grown too delicate.
I grieve thy black skeleton,
Still overlayed by fate.

O could I delve thy shimmer,
Such forth thy shiver, and for
All time clasp the answer,
What protection, who the protector.

'I WENT TO SEE IRVING BABBITT'

I went to see Irving Babbitt
In the Eighteenth Century clean and neat
When he opened his mouth to speak French
I fell clean off my seat.

He spoke it not fair and fetisly
But harshly laboured it like a Yankee

Even as my nubian Swahili
Is sweet and pleasant to me.

And when we went out of the critical door
Crying for more, crying for more
I saw the hater of mechanical America
Bulge through the Square in a critical Ford.

Harvard is a good place, Harvard is the best,
Among the immemorial elms you'll come to rest
Strolling the Yard, the only proper yardstick,
Warbling your native foot-notes mild.

RECOLLECTION OF CHILDHOOD

O the jungle is beautiful the jungle is wild
Here are the rodents and the butterflies,
The thorn tree prickles and the shady grottoes,
And I'll lie in the sun all day, and the shade.

And here is the orchard, bulging big fruit,
Adazzle in the eyes all red and golden,
There is the barn, and the many cows,
Hurray for the hay, I'll fall and tumble in the hay.

And below the happy pasture green and moist
I'll walk it to the shady grove by the river,
Swim the clear stream, the chalk cliff climb,
Bursting with desire still—cease action never,

But in the natural world of happy forms
Bird, beast, tree, fern, earth and rain and sun
Live melodious love, and touch above
All fingered air the very god of love.

ORCHARD

I

Lovely were the fruit trees in the evening.
We sat in the automobile all five of us,
Full of the silence of deep grieving,
For tragedy stalked among the fruit trees.

Strongest was the father, of solid years,
Who set his jaw against the coming winter,
Pure, hard, strong, and infinitely gentle
For the worst that evil brings can only kill us.

Most glorious was the mother, beautiful
Who in the middle course of life was stalked
By the stark shape of malignant disease,
And her face was holy white like all desire.

And we three, in our benumbing youngness,
Half afraid to guess at the danger there,
Looked in stillness at the glowing fruit trees,
While tumultuous passions raged in the air.

II

And the first, the father, with indomitable will
Strove in iron decision, in all human strength
With a powerful complete contempt of defeat,
Six feet of manhood and not a mark of fear.

And the next, the mother, wonderfully mild,
Wise with the wisdom that never changes,
Poured forth her love divinely magnified
We knew not by what imminent despair.

While the older brother and the younger,
Separate, yet placed in the first light

45

Of brutal recognition, held a trembling sister
Who knew not the trial of fortitude to come.

And in the evening, among the warm fruit trees
All of life and all of death were there,
Of pain unto death, of struggle to endure,
And the strong right of human love was there.

THE SOUL LONGS TO RETURN
WHENCE IT CAME

I drove up to the graveyard, which
Used to frighten me as a boy,
When I walked down the river past it,
And evening was coming on. I'd make sure
I came home from the woods early enough.
I drove in, I found to the place, I
Left the motor running. My eyes hurried,
To recognize the great oak tree
On the little slope, among the stones.
It was a high day, a crisp day,
The cleanest kind of Autumn day,
With brisk intoxicating air, a
Little wind that frisked, yet there was
Old age in the atmosphere, nostalgia,
The subtle heaviness of the Fall.
I stilled the motor. I walked a few paces;
It was good, the tree; the friendliness of it.
I touched it, I thought of the roots;
They would have pierced her seven years.
O all peoples! O mighty shadows!
My eyes opened along the avenue
Of tombstones, the common land of death.
Humiliation of all loves lost,
That might have had full meaning in any
Plot of ground, come, hear the silence,
See the quivering light. My mind worked

Almost imperceptibly, I
In the command, I the wilful ponderer.
I must have stood silent and thoughtful
There. A host of dry leaves
Danced on the ground in the wind.
They startled, they curved up from the ground,
There was a dry rustling, rattling.
The sun was motionless and brittle.
I felt the blood darken in my cheeks
And burn. Like running. My eyes
Telescoped on decay, I out of command.
Fear, tenderness, they seized me.
My eyes were hot, I dared not look
At the leaves. A pagan urge swept me.
Multitudes, O multitudes in one.
The urge of the earth, the titan
Wild and primitive lust, fused
On the ground of her grave.
I was a being of feeling alone.
I flung myself down on the earth
Full length on the great earth, full length,
I wept out the dark load of human love.
In pagan adoration I adored her.
I felt the actual earth of her.
Victor and victim of humility,
I closed in the wordless ecstasy
Of mystery: where there is no thought
But feeling lost in itself forever,
Profound, remote, immediate, and calm.
Frightened, I stood up, I looked about
Suspiciously, hurriedly (a rustling),
As if the sun, the air, the trees
Were human, might not understand.
I drew breath, it made a sound,
I stepped gingerly away. Then
The mind came like a fire, it
Tortured man, I thought of madness.

The mind will not accept the blood.
The sun and sky, the trees and grasses,
And the whispering leaves, took on
Their usual characters. I went away,
Slowly, tingling, elated, saying, saying
Mother, Great Being, O Source of Life
To whom in wisdom we return,
Accept this humble servant evermore.

GRAVE PIECE

The lustral sweat in its fine slow beads,
Loosed from the massive burning brow,

>   (I try something new, I break my skull.
>   Death, I try to get into you.)

Bit drills, bit little axes, like heavenly saws
Through sunshine, incrassated on the stone,

>   (I make a picture-map a process,
>   For the instruction of logicians.)

The glassy lapses of uncloven time.
In, tender; in, bitter; in, indifferent,
The shiny spheres their nakedness assail,
Tiny wanton world revelling joyers,

>   (Death comes in the whirlwind, or like a cat,
>   Death creeps down like a suspiration,
>   Because I have life Death is Death
>   And through Death I try to reach perfection.)

Seepers; and spread a frostwork hands
There in the vasy tomb of bone-green icicles,
And gelid grasses of cold bruise to touch;
Of looking roots, night agate eyed; and

48

(Love is in me, love Protean
Strange and potent, as the stone uncloven.
And all the sunny afternoon of summertime
Think, wink, entombed, feel, steal)

Lynx-lensed weeds' fixed-pricking stare,
And air with motion scarce as sight
Above born squinting flowers, screaming, frozen,
Composed in a scenic breakage of blear light

(Love flooding me, I would overcome the world
I must discover inexorable Death,
Death and Imagination conjoining—
—tickling gnat my fingers flies from)

To static points and rays of endless sameness,
The things themselves, their fatal natures;
And the syzygy caused a stony sound
Through hollow, skull-shaped geology,

(Attempting to do what cannot be done,
Poetry to break the marble word
Even as time sullied the carven name,
And I am trying to fly through Death like a bird.)

And from this I saw grow
Although my sense shook with fear
A crystal Tear
Whose centre is spiritual love.

THE HUMANIST

Hunting for the truly human
I looked for the true man
And saw an ape at the fair
With the circle still to square. Learning
Breeds its own ignorance. Fame and power

Demand a rush and pounding.
Those who rush through rush through, and who
Are they but those who rush through?
Yet truth resides in contemplation
And comprehension of contemplation
Not necessarily of Plato. Action explains
A field full of folk and golden football flexions,
Action for actors. Truth through contemplation
Resurrects the truly human,
Makes known the true man
Whom these lines can scan:
But miss his secret final point.
The world is too much in joint. No use
Setting that right, that squaring, that harrying: for
The true man lives in mystery
Of God; God his agile soul will see
But he will not see God's majesty,
And that is what makes you and me
Whether man or woman
Neither true nor free
But truly human.

THE VIRGIN

I will not think of her in her coffin
With worms' teeth needling a death's head
In her red, and white, and soft cheek;
Her resilient hair, crushed dead,
That only the boy winds fingered.
The virginal nipples pinched by stones.
The silver belly sullied by slime,
Kiss unreciprocate; bones, O!
Without strength to round the womb.
And April skin-shine browned and fouled.
Stinking under the sweating earth,
The mind's whole marble disembowled.
Save me these pictures, you denisens,

50

Captain worms of the gross earth,
That I may tear from my mind such album,
Formal decay: death's hideous birth.
No better than a beggar, a base
Whore, or baser, a designing lecher;
Equal to Helen, diced with jewels,
Putrid at the bawdy event, like her.
My poor Moll, dead as a stump,
It's good I knew you not, sweet,
But in feeling's jammed proximity;
Doltish memory, cranky perception
Give you the lie, for my brain sue,
While, mured up in this ruinous vision,
My mind beats lively as a disease.

'MAN'S GREED AND ENVY ARE SO GREAT'

Man's greed and envy are so great
He'll go to any length or date
To beat down his fellow man,
If beat him down he can.

Though under dread he cower,
He'll do it in the name of power
That obvious, that visible success
Prove to the world his canniness.

And let Christ go; let Truth
Hang; praise the fang and the tooth,
Force life though like a balling youth,
The truant animal in action.

Praise to the thinkers and the dreamers
Those dear quaint saintly creatures.
They evade the blood to build the brain,
Philosophic men who love the sun.

51

'THE GOAL OF INTELLECTUAL MAN'

The goal of intellectual man
Striving to do what he can
To bring down out of uncreated light
Illumination to our night

Is not possession of the fire
Annihilation of his own desire
To the source a secret soaring
And all his self outpouring

Nor is it an imageless place
Wherein there is no human face
Nor laws, nor hierarchies, nor dooms
And only the cold weight of the tomb

But it is human love, love
Concrete, specific, in a natural move
Gathering goodness, it is free
In the blood as in the mind's harmony,

It is love discoverable here
Difficult, dangerous, pure, clear,
The truth of the positive hour
Composing all of human power.

'IF I COULD ONLY LIVE AT THE PITCH THAT IS NEAR MADNESS'

If I could only live at the pitch that is near madness
When everything is as it was in my childhood
Violent, vivid, and of infinite possibility:
That the sun and the moon broke over my head.

Then I cast time out of the trees and fields,
Then I stood immaculate in the Ego;
Then I eyed the world with all delight,
Reality was the perfection of my sight.

And time has big handles on the hands,
Fields and trees a way of being themselves.
I saw battalions of the race of mankind
Standing stolid, demanding a moral answer.

I gave the moral answer and I died
And into a realm of complexity came
Where nothing is possible but necessity
And the truth wailing there like a red babe.

'I WALKED OUT TO THE GRAVEYARD TO SEE THE DEAD'

I walked out to the graveyard to see the dead
The iron gates were locked, I couldn't get in,
A golden pheasant on the dark fir boughs
Looked with fearful method at the sunset,

Said I, Sir bird, wink no more at me
I have had enough of my dark eye-smarting,
I cannot adore you, nor do I praise you,
But assign you to the rafters of Montaigne.

Who talks with the Absolute salutes a Shadow,
Who seeks himself shall lose himself;
And the golden pheasants are no help
And action must be learned from love of man.

A MEDITATION

Now you are holding my skull in your hand.
I am not anybody. I am dead.
God has taken my life away. You are holding
My skull in your hand, the wind is blowing
(Blowing, the ineffable structure of the wind)
The wind is blowing through my skull like a horn
And you are thinking of the world's unearthly music
And of the beauty beyond the earth and seas,
You think the tune will tell you of the truth,
You are thinking: thinking, are you?—are you
Thinking? I am dead. I am dead and in your hand.

Now what does it matter whether I lived or died
Or when I lived or when I died or what I tried
To do of all the things there are to do in the world,
You might be musing upon a finch, or a melon,
You might be not you but somebody else, some other,
Some utter muser, maybe, in some other land
With some other kind of brains or bones, or both,
Some lack of bile to make these peregrinations;
You might be a man who never found me at all,
That miraculous fellow, that idoitic genius
Who never had to think of the wind in my skull at all.

There, you see I am so helpless in your hands.
I cannot get back, cannot reach or yearn back,
Nor summon love enough, nor the intellectual care—
Being dead, you talk as if I had spirit at all—
To come back to you and tell you who I am.
You do me too much honour with your grave words

And quizzical head bent down, trace of a lovable smile,
Too much too much honour for one so windless,
And witless, wizened long past wandering and pondering.
Go back to your strict duties of the earth, man,
Make love to your girl long nights, and long summer days.

But there, you hold me, a philosopher! If philosophical,
You had better put me back in the earth,
Hold nothing at all! Hold your own hand, man!
Now you can feel the dainty wind playing through it
Weaving among your fingers what you want to know.
There is the place to love and make your contemplations
For you can still make your fingers fly and go.
Of all symbols the cunning mind devises
None can be so strict, so purely right
As the hand that makes the fingers fashion the love
Contending in immaterial mastery.

But something touches you, deeply, beyond all comment,
Something so tender, wise, human; you are only human
And grief makes you as a little child that weeps
For you do not know why you are in the world
Nor how you got here, in any real sense, nor how long
Before some ruthless shock will end you forever,
That peculiar you you must cherish, the uniqueness,
The kind not like the skull that is in your hand
But that is fixed in its intricate nicety,—
As who can imagine, seeing a fine bull in pasture
That that magnificent creature will lie down and die?

God has taken my life away. I am dead,
I am not anybody. I am no oracle. You be
An oracle, if you can. You be full of imagination.
You see beyond me, which is seeing beyond yourself,
Yours is the purely human burden and the prophecy.
These things can only be talked of by men.
Man is the talking machine; or the arcane marionette

Who like a god thinks himself to be immune
From the terrible void and absolute darkness
The unthinkable loss and final destitution
Upon which he builds (O! yes!) his mighty hierarchies.

But if I had the power to seem to be,
Or could even wish to be powerful, and say
'I am Lord of Life. I am Eternal Life'—
I could not say it, the contradiction is too great.
This goes too far, and lacks a certain humour.
You build a fiction of what I am, or could be,
And amidst the limits of your mentality and words
Revolve, with all your feelings stumbling blocks
And your keenest thoughts barbs of fiery wire.
O dear sweet little timeful poor creature
Tell this skull what it cares not to say

Life blows like the wind away

You have run the weirs of the emotions,
Even as I did; but they are not through with you yet.
There will be many complicated messages,
Memory will make of your heart a veritable mill,
The need for action jerk your legs, ambition
Choke you, rage consume you almost to inanition,
Conquest will thrill you, and defeat annul you.
And you will not have enough, but lust of life
Drive you on to the cold, very brink of the grave
With man's old captive, cyclic wish to know
What it is all about, meaning and moral dimension.

What is it all about, you are asking me now
For it is less of me than of yourself you are thinking,
Surely more of yourself, flesh-bound as you are—
You would withdraw in horror at my secret,
You would not want to know, your long-lashed eyes aglare,
Of the cold absolute blankness and fate of death,

Of the depths of being beyond all words to say,
Of your profound or of the world's destiny,
Of the mind of God, rising like a mighty fire
Pure and calm beyond all mortal instances
Magnificent, eternal, Everlasting, sweet and mild.

No, I rise a little, I come to life a little.
If I could only make you see the simple truth,
So simple it is so hard to say, hard to believe,
That you are to be man, that is, to be human,
You are imperfect, will never know perfection,
You must strive, but the goal will recede forever,
That you must do what the great poets and the sages say,
Obeying scripture even in the rotten times,
That you must, also, not think nor feel too much
(For the Greeks knew what they were talking about),
And that, as you have lived, so must you die

And blow like the wind away

Now you walk back among the flowers of the world.
You have put me down and I am myself again.
Now you return, return from your solemn meditation
Among the strange shadows, the strange veils of longing
And through the cathartic action of beautiful contemplation
Seek among your fellow creatures whatever is good in life,
Purified by this; for, as sin purifies lust,
As war purges society, as death rarifies life,
So the contemplation of death is valuable
Restorative of the soul to new even reaches
Easing a little the burden of our suffering

Before we blow like the wind away

And blow like the wind away

'THE FULL OF JOY DO NOT KNOW; THEY NEED NOT'

The full of joy do not know; they need not
Know. Nothing is reconciled.
They flash the light of Heaven indeed.
Let them have it, let them have it, it is mild.

Those who suffer see the truth; it has
Murderous edges. They never avert
The gaze of calculation one degree,
But they are hurt, they are hurt, they are hurt.

RUMINATION

When I can hold a stone within my hand
And feel time make it sand and soil, and see
The roots of living things grow in this land,
Pushing between my fingers flower and tree,
Then I shall be as wise as death,
For death has done this and he will
Do this to me, and blow his breath
To fire my clay, when I am still.

'COVER ME OVER'

Cover me over, clover;
Cover me over, grass.
The mellow day is over
And there is night to pass.

Green arms about my head,
Green fingers on my hands.
Earth has no quieter bed
In all her quiet lands.

THE RECAPITULATION

Not through the rational mind,
But by elation coming to me
Sometimes, I am sure
Death is but a door.

A state of purity, sweet grace
It is, nor can last long,
But in that essence, I feel
Life beyond death is real.

Perhaps it is only the human need.
When reason rules, reason denies it.
But comes elation unto me
And blows God all through me.

When the flesh was young and strong
And the mind also, I denied God.
But time's sullen call
Has made my body fall.

Spirit of holy love, arise,
Meek and gentle, sweet and calm,
Arise in the ruthless world
And your truth put on.

To the rational mind grant
Things rational. But to the spirit
(All things there possible)
Accord what is spiritual.

'IMAGINING HOW IT WOULD BE
TO BE DEAD'

Imagining how it would be to be dead,
Until my tender filaments
From mere threads air have become
And this is all my consciousness

(While like a world of rock and stone
My body cumbersome and big
Breathes out a vivid atmosphere
Which is the touchless breach of the air)
I lost my head, and could not hold
Either my hands together or my heart
But was so sentient a being
I seemed to break time apart
And thus became all things air can touch
Or could touch could it touch all things,
And this was an embrace most dear,
Final, complete, a flying without wings.
From being bound to one poor skull,
And that surrounded by one earth,
And the earth in one universe forced,
And that chained to some larger gear,
I was the air, I was the air,
And then I pressed on eye and cheek
The sightless hinges of eternity
That make the whole world creak.

'I WALKED OVER THE GRAVE OF HENRY JAMES'

I walked over the grave of Henry James
But recently, and one eye kept the dry stone.
The other leaned on boys at games away,
My soul was balanced in my body cold.

I am one of those prodigals of hell
Whom ten years have seen cram with battle;
Returns to what he canted from, grants it good,
As asthma makes itself a new resolution.

I crushed a knob of earth between my fingers,
This is a very ordinary experience.
A name may be glorious but death is death,
I thought, and took a street-car back to Harvard Square.

THE INEFFABLE

When Eve ate the apple
My woes began,
But I didn't believe then
That I would believe in Original Sin.
When Adam delved
I began to fail,
But I didn't comprehend
The majesty of the Supra Rational.

That Cain and Abel
Should be such brothers
And not love each other
Seemed not in the nature of things.
These were old fables
Of the dimly understood,
When man began to construct
Evil and Good.

Than the passion of pied air
Flew through the days,
The days used the years,
The years grew to tears—
Logic made the stories go
And imagination moved the key.
Somewhere Love came in,
Quietly, mysteriously.

The force as odd as any
After all is said
Has more sense in it
Than hounding the dead.
And the bond of the obvious
Blood in its beauty
Controls more, is not moral,
Simply is.

When nothing can be understood,
World-systems crumble,
Love is of the essence,
Poetically speaking.
As in music
The truth is in between the tones,
Neither action nor will
Tell what is going on;
It is something glancing off
Allures always,
The rich sense
Of the all, the impossible, the ineffable.

'MYSTICISM HAS NOT THE PATIENCE TO WAIT FOR GOD'S REVELATION'

KIERKEGAARD

But to reach the archimedean point
Was all my steadfastness;
The disjointed times to teach
Courage from what is dreadful.

It was the glimpses in the lightning
Made me a sage, but made me say
No word to make another fight,
My own fighting heart full of dismay.

Spirit, soul, and fire are reached!
And springs of the mind, like springs of the feet
Tell all, all know, nothing wavers there!
All the flowers of the heart turn to ice-flowers,

Heaviness of the world prevailing
('The higher we go the more terrible it is')
Duplicity of man, heart-hate,
The hypocrite, the vain, the whipper, the cheat,

The eternal ape on the leash,
Drawing us down to faith,
Which the Greeks call divine folly,
The tug of laughter and of irony.

THE DREAM

In a dream I was in a tunnel struggling
A chalky tunnel, and torturous place,
With at the end a suggestion of light,
A space of light and gruff old optimistic strokes
For though convulsed, frisked, I felt I would reach the light

And as in a dream things happen without pain,
In a childhood sequence of untractable reality,
In this pleasurable though unpredictable predicament
I struggled, drawn directly by the great light,
The tunnel was merely there, exfoliating.

Then as if angels whisked you, an unseen force
Yet seemingly not a force, for all was fervent balance,
I was in high regions of beautiful world and life,
I visited the extreme palaces, stroked the glowing air,
Went up through hitched forests to a gold plateau

And all was triumph, magnitude, deep vistas,
All was largess of harmonies, freedom of form,
Luminous periodicities in a static realm,
The unimaginable godhead, divine peculiarity,
The child's, the death's-head's unconquerable vanity.

THE MOMENT OF VISION

The moment of vision will come when it will,
The gift of the gods, the unshakable
Unbreakable
Subtle intuitive stewardship, and still
That matchless satisfaction of the maker.

To read the spirit was all my care and is,
To lose life to find hope,
To destroy in the heady grope
For new worlds upon an old's effacement,
Through the breach of evil the breath of grace.

Those men who fight for the spirit have my love,
Fierce young poets tearing their hearts out,
Strong for many materialistic bouts,
Destroyers who put on creation like a glove,
Wear grace from their despair: the wrenched fist.

The worthy one who did a worthier thing
Than all the worthies did—he kept that hid;
Him whom another age heard not,
But now my surest music of the word,
Earth-thrown singer, who burnt his bridges.

RETROSPECTIVE FORELOOK

The logical pleasure of expectant mercy
Will be an end of us, will make an end.
The demon pulling us from the peach orchard
Where we were uninfected, ineffectual in time,
Cannot get off our skins; we did not win,
But entered more fully the manhood of sin.

The book of Boehme, being tired after bicycling,
The visionary voluptuary asleep in the hay,
All that effort in the fresh fields of the French,
These were days were prayers were potent maskers,
Fame on his forehead, the gust a good gear,
Where now the glory of the ghostly year?

Pascal, too, the attic full of strife and charm,
Mozart a checkerwork of lilacs; Descartes,
Who in his statue, strangely singular in Tours,
Set the savage mirror in the delving eye,

These were pilgrims come to country new
To pommel, hearten, sing, seem, and sue.

The old vexation of the moon! Climbing Chartres,
Where on the windy roof, midst gargoyles intricate,
What moonlight and what noonday imaginings
Compelled the reason and made dim the blood
In shadows of the past over the champaign
In silent noonlong or roan night of rain.

While in that vault where mostly man makes God,
The hallowed light at evening dimly falls,
(the dimness of the nave gravely the light recalls)
But falling brings the glow and golden growth
Of vague, supernatural, spiritual pressures
where the weight of our redemption seems
A compilation of and nurse of dreams.

And long were days and tall were tremble-trees,
The symbol of the self lost in a haze.
And errors of sentience, O senses sweet! sour! sweet!
That ravelled out your substance quizzically,
To go forward and to retreat, to gain
And lose, and never to quite see the truth plain!

That struggle of all Europe! The massive English means
Contiguous to reason, lapsed in jiggly greens,
Where heart was pierced, harrowed the head, was bride
To lust of learning, and the whole eye acquist.
There were the centuries a valid looks,
On English faces no need to read the books.

The demon drives us and we have to go.
The dangerous mountains never climbed aspired to,
While tops of standing, dancing places, are descried.
Consciousness it is that blights us all,
Visionary candor in the branches
Hallucinatory avalanche the pandar.

While stealing through the nave so ancient seeming
The truth of metaphysical reason there concordant,
Stark Lenin, radiant, with blood upon his head
And hands like hams that fed a famished land
Was all that subtlety is worth, and more,
Was the West, was what the world would be.

THE LYRIC ABSOLUTE

I

Art, the holy care of life
Excruciate, and the wife
Of love, in sounding strangeness sleeps
Through the wars of the active deep,
And I hear the struggles and wails of desire
Sunk in redemption's purifying fire,
And I am made like a child
Who would not be mild.

II

Images of death arise
And scatter the skies.
Images of life depend
Upon joy in man.
The stars of day fall
And the night arises bleak
And Christ in whole blood
His Child in Man seeks.

'I WILL NOT DARE TO ASK ONE QUESTION'

I will not dare to ask one question.
Fondling the warm sun while I may,
A man married to oblivion
I will kiss her mistress lips all day;
But in this bed the solemn blooms
Of the darkness grow in the bones,
That are a slow drug to the mind.

I will not dare to ask the stones.
One night alone told quietly
How I must move back to the womb
Which nourished me on the unknown
And once more in that love entomb,
Thus, without answer, simply,
Gone, damned, the loam over me.

NEW HAMPSHIRE, FEBRUARY

Nature had made them hide in crevices,
Two wasps so cold they looked like bark.
Why I do not know, but I took them
And I put them
In a metal pan, both day and dark.

Like God touching his finger to Adam
I felt, and thought of Michaelangelo,
For whenever I breathed on them,
The slightest breath,
They leaped, and preened as if to go.

My breath controlled them always quite.
More sensitive than electric sparks
They came into life
Or they withdrew to ice,
While I watched, suspending remarks.

Then one in a blind career got out,
And fell to the kitchen floor. I
Crushed him with my cold ski boot,
By accident. The other
Had not the wit to try or die.

And so the other is still my pet.
The moral of this is plain.
But I will shirk it.
You will not like it. And
God does not live to explain.

TRIPTYCH

PERCY

Whimsicality is a thing in itself.
Cast out your noble argument.
You've only got to go look
At a child fetched forth from the womb,
Or see an old codger croak,
To assume the world is nothing but sorrow.

JOHN

But, you forget, sir,
The importance of values,
Of rigorous discernment among them.

PERCY

Ha, your cheeks will be long
Enough to make sizable culverts
In them, down which to flush
The tearful sorrows of your seriousness.
Come, praise whimsicality.
He's a painted clown can walk a
Tight rope narrowly, and will fool the people.
He's your only modern who can't be criticized.
He's understudy to the true bigwigs,
Those who experience experience.
Mark his moony jaws and craven
Pate, his withal constancy.
Critics be brought to the tent.
They have not learnt all.
They want juice.

JOHN

My dear sir,
You run on like an adolescent litterateur.
Come out of it.
Let us get down to reality—

Ho, reality is up. What's up? So
We are up to reality. I
Insist on levity. There's a
Certain deviltry in all levity.
Let your philosophical discourse
Wait on the slow decay of paper.
The shelves will last and the books on them,
Till the digger worms set up a Piccadilly
And haul stuffs hard through the tubes
That the forage of nature's
Dumped atop the sheaves of art.
Spirit won't keep.
But like the laughter in the throat
It shocks out over the teeth
In a stony springing brook;
Falls over an edge and
Becomes an airy gauze;
Becomes great laughing-crystals.
The worms have no guts,
And must eat only wood and paper.
I am Air-eater,
The hardest, rarest occupation.

JOHN

It is time I praised my own tolerance,
But I have known you so long
I will allow you the pleasures of expansion.
However, I have for the past fortnight
Examined the five volumes of the late Dr. Fardel,
The latinist we knew at the University—

PERCY

If you
Would put all the great brains of
Oxford and Cambridge in a tub
And boil them for three and thirty minutes,
You might beget an elf.

JOHN

As I was saying, the learned Doctor—

PERCY

Oh—I feel a pun coming.
Hmmm. A gross pun, although
All men hate them, is worth more than
A quire of your syllogistic Doctors.
Is an integer of sense, a compound
Of fertility; albeit an indigestible squib.

JOHN

Spare me your puns anyway.
Now, his theory in general is—

PERCY

I wonder if he theorized in his lady's bed.
No doubt, being a Latin scholar,
He went over old conjugations.

JOHN

This so-called merriment of yours
Is merely dull superstructure
You erect upon your ignorance.

PERCY

If I'm ignorant, then I love ignorance.
I'll squeeze it into a voluptuous guffaw then.
And I'll shout ignorance at you
Until you get some sense.

JOHN

Such a belligerent attitude is hardly
Becoming to one of your formal education.
But in our last discussion, I recall
We had seriously got to the point of—

The point of merriment at you, sir!
When a man is out of work,
And loitering in his father's plush arm-chair,
It's grand to stalk the realistic pessimists.
But I am for the point of ignition:
Let the intellectual hayricks
In the wet season be burnt up by
The sun's hilarious eye.
Asthma has got a grip on existence.
That is my particular gripe. A jape.
In truth, I am so full of glee I am
More ridiculous than before I said so.
Come, now, have you ever seen
A critic jump a fence?
'Twould be damned bad for his criticism.
I'll bet the barbed wire would
Criticize his backside, make him
Bleed out his petty principles;
He should have jumped higher, and got over the Greeks.

JOHN

Well, I guess I must leave you
And go back to my discourses.
You seem incapable of
Even an attempt at lucidity.

PERCY

What, lucidity?
When Lucy is lucid to me
I will tempt her and attempt her
And run and tell thee.
Pardon, I retch. It must be the acidity.

JOHN

You make me sick.

PERCY

Be merry. As long as you are not married,
You might as well be merry.
Really, tomorrow I will cry all day.
I will buy little squirt guns filled
With sugar water and run them down
The two lobes of my face,
To let you know I am crying.
Tomorrow I will be the soul of solemnity.
I will go to the Museum and lie down
Full length in a Medieval stone coffin.
I did it once, and was so grievous a sight,
The attendant broke out laughing.
Let me groan from six o'clock in the morning.
By noon I will have got the habit,
And by night I shall be an absolute wailing ghost.

JOHN

Come, you are not reasonable.

PERCY

Reason and treason are the same thing to me.
Reason is too credulous.
It irks me like old ladies;
Or old gubernatorial gentlemen.
Ha! I fly by Absolute Force.
I am so light the air is my very being
And I would be all over the sky.
Your reasoners only poke about in old clothes.

JOHN

You simply rave against what you love.

PERCY

If I rave against what I love,
You fail me in reciprocal

Love of what you hate.
Oh, there comes a she.
Now I must fly against
The earthly quality of women.
Hello, vegetable, how's your green hue?

What sport is this?

Good afternoon, dear. My greetings. Don't bother
About him; he's in a ripe inconsequence,
And won't bother us a bit. What a lovely frock—

I have just come from town with a car
Full of sheer things, the smartest things from Paris.
And these matching jewels from India.

Oh, for a new attitude to death.
The Polynesians did well when they
Packed and squeezed the head down to
The round and look of a sick lemon
Left over long in the sun. Fond embalmers.
The bulging fellow got down at last,
Brains confessed in a boy's handful.

What an awful fool. How should I
Provoke such language, and with my new ensemble?

As host, I must ameliorate
Between your displeasure and his candour.

73

I propose tombstones and mausoleums
Be laid under the pile-driver
And done into powder for the cheeks of virgins.
Such quantities of the stuff would be,
And so few authoritative patronizers,
They'd be chalked to the death withal,
Tricked out with the most emaciated spells
To get their man. Now, these dead.
What to do with the dead. I once saw a
Groundhog stick his fat corpse
Into the very light of July.
It blistered, broke loose, and you've never seen
So merry a sight. Soon he was aswim
In his own oils, hot as a lecher,
And then he was dancing with maggots.
In a fortnight he was itching with
A most greedy dissipation,
As hilarious as a bunch of bees,
And strong enough to drive off a troop of buzzards.
By August middle he had got strained
Into a nervous maniacal gymnastic,
And by December his divine frenzy
Was quits with the world.
I passed by the place the next summer,
It had a lonely look.
Indeed, he had only a little hairy ghost.
But it would boil again; and manure the wild flowers.
Now take corpses, stretch them out
In bold capable nakedness
On the ground and see them do likewise.
And liveliness to them!
What a purgative performance.
'Twould make people Buddhistic,
And do away with the wars.
You'd have a better opinion of life
For the stench. Nothing like strong youth,

Strong love, strong corpses boiling,
Quick with effervescence.
And hang on their heaving breasts
The jewels of women. They'll sink
Down slowly in the fleshly mess,
And at the last shine on and dazzle the earth.
An acute eye will pluck the glory
Of their indestructibility.
Page Hieronymus Bosch.

JOHN

You betray yourself, fellow,
And in the name of frivolity
Entertain us with a quite serious notion.
I glimpse even a spiritual
Regenerative principle
In your suggested and unlikely tactic.
But you employ a symbol for your
Metaphysics; otherwise they would vanish.
They would go back to the blood
And become the simple energy of action.
There is no necessity to change
The various burial rites of the world;
They are forms of human decency.
You would do better to apply your fancy
To the living,
Who stand in need of vitalizing notions.

PERCY

Have at you. I am Air-engine.
And incandescence is the sole sweet of the world.

JOHN

You are in danger of becoming a theolog an.
If you would put your views
Into a somewhat less heretical mould,
Even encase them in a cloak

Foreign to their character,
You might get a Rockefeller fellowship,
And study a year at Rio.
But you are only man after all,
The actor-animal.

Well, apparently I am here
Only as an objective symbol of
Your opinion of women:
I must suffer your talk.
You are always dissatisfied;
Fomenting ideas
As if the world could be changed.

PERCY

There's no more happiness in the world
Than a squirrel has with a nut.
You crack it open, eat it up,
And want more. But ideas
Are everywhere in the sky.
Come, fly by whimsicality.

JOHN

You ought to work for the City Company.
You could wear spikes,
Climb up poles like your ancestors,
And thereby carry electricity to the people.
If you were less ubiquitous and more single
I would want to cope with you.
Come, let me put some salt on your tail.
Your shallowness—

PERCY

Shallow? Then air is shallow,
Through which we see to heaven.

Shallow? Then water is shallow,
Of which we are all composed.
Shallow? Then morning's atmosphere,
Lakes, rivers, rills and streams
Are shallow. What is your jealous depth
But layers and layers of shallows? What
Profundity, but a schoolmaster's
Multiplications of a feather falling.
Shallow? Then I'll lie down on the wee
Boat of a feather, and sail in the heavenly air.
I count on the docility
Of men to save them much trouble.
Here is an old one, a seller of papers,
Bearing his load of defunct news
Homeward in the centre of night.
Life makes him
Faithful to his duty. He says
Good-night gently, when you pass.
And I daresay he is as good as
Your schizophrenetic synthetic dictators.
Shallow? Come praise whimsicality.

PRISCILLA

I say, do you mind, I must be leaving——

JOHN

Why, I'm awfully sorry, won't you——

PRISCILLA

Good-bye——

JOHN

You've driven the lovely creature off
With your frail bombast.

PERCY

I must go. This is not what I was.
This is something else. This is pure Phoenix.
Come feel it. No, you can't. For though
You were vampire and drank my blood,
You would not get me. Fetch me
The past in a basket lined with grass,
If you think to. Bring me
A bucket full of huckleberries
Plucked at ten years old, if you will.
Present me my first virgin,
If you can. Or dig up my bones
When I shall have been a decade dead,
If you are able. No, no,
I am something else than I am.
I am the tickings of clocks whose ratchets
Wear away now in the mill stream.
And the rust of them, standing in old barns.
I am the bud of the leaf
Whose remains stick on my heels . . .
Oh, I am the lad's eyes,
When there was wonder upon them
Like flax twinkling in the wind and sun . . .
The same skull; Plato and Aristotle,
Plus Shakespeare, and Donne, and Blake, and Marx,
And as gone as old Neanderthal.

JOHN

Would you mind coming along?

PERCY

Oh, that's all right, but the world's
Nothing but a pile of filth. The
Relations between men are as excellent as hell,
They are all well ordered and arranged beforehand.
There's nothing so full of end-results

As friendship; and nothing so crafty
As men in love with mountains.
When you have gone through all,
Evil stares you in the strict, emaciated face.
But you're not emancipated yet.
You can no more conquer yourself
Than walk on Pluto. The good falls
Into a big disrespect and a dusty bin;
Evil flowers out incandescent,
Ready to burn you up again.
The masters of life have as much
Energy as it; the rest fade off.

JOHN

I say, do you mind coming along?

PERCY

There are those who say flowers
Have no more to say; look there,
There is the world pure and final,
Nothing else to it. There are those
Who say those are partial fellows,
Befriended by lenses. These are troubled
By a fire in the guts and say
Microscopes do no good.
The messy little bluet is only
A sallying forth of some knotted ganglia.
They have a great room between the ears,
Wherein they play games. There are others,
Who can hardly tell whether they are
In Venice or Canton, but they are
Sure neither has the importance of a thumb tack.
They doubt they doubt their own doubts,
And end where they began—a right
Doleful dingle-dangle state indeed.
The experimenters are, sooth-la,

The only realizers. That is
Because they refuse to think.
They put chemicals in test tubes,
Measure the results, and have no
Astonishment whatever at the conflagration.
But the artist saves the world.
I have been everywhere, done everything.
There is nothing left but
To search out the bosom of God,
The chief receding pleasure known to man.

The searching is ever better
Than the finding; when you have found something
You must keep it, which is another kind of searching.
If you search for an unknown woman,
You must always be seeking. If you
Marry one, you must, though you have her by you,
Seek out new ways continually to appease her.
Get a child, and you go back to God.

JOHN

Well, come along now, Percy.

PERCY

I have come to no conclusion.

JOHN

You have come to no conclusion.

PERCY–JOHN

We have come to no conclusion.

PERCY

Conclusion is too inclusive.
To(o) crude warfare my food is air.

ODE TO THE CHINESE PAPER SNAKE

I

Held on the slightest of bamboo poles,
Suspended from the most voluble of delicate strings,
Heavy in the head, as is proper to subtelty,

His limitation is his slyest and activist, most charm;
Achieves with the least energy the greatest purpose;
Frail in the body, liable to tear, a dissolute.

His eye is horny, he has the feather of imagination
For a fang; his hiss is what you are.
The circuitous pleases him most: circles are endless.

II

The father of a metaphysical principle,
Made in Hong Kong but sold in San Francisco,
The paper snake is foreign to the people of Maine
And has never been seen in the hives of Vienna.

Nearest to stillness his force is most potent.
Thus he adumbrates the meaning of innocence,
Liable in gross action to the unsalvageable and comic;
His least motions manipulate the fiercest evil.

Delicacy is also the mark of civilization
And this is a most uncultured Satanism;
Him no Spengler fingered into exotic play,
Nor divested him of his striking parody.

III

As sex he would be simple, but nothing is simple,
So sexually he bares the fascination of centuries.
He triumphs adept, formally, evident, exquisite.

The ultimate investing philosopher of the future,
A paper snake gives way to none in egoism.
His confidence is masterful and ritualistic,

His is the innate totality of knowledge.
A pertness poses his profundity, deftly expressed,
And it has in it the guile and the smile of the sly.

IV

The pleasantest and the most profound trickery
Is that you can handle as you will this apostle.
It is you who think you control the principle of evil,
It is you who think you invite the charming play of the good.

Nor should we forget forgetfulness, which he induces;
Such rite of resolution is another of his provinces.
Neither Chinese nor Christian, his characteristic of universality

Intrigues as blood, as intellect, as purity, as impurity;
Most inanimate the Chinese snake becomes most animate
Seemingly in the very laughter and sorrow of your being.

V

Aeschylus should have asked of this plaything an answer
As Sophocles spent a lifetime in honouring him;
He was too inward for the Western tone of Shakespeare;
Baudelaire knew him in Paris as a cat.

The heaviness in the Hebrews did not see his dandyism.
Perhaps the greatest effrontery to knowledge
Is that you can manipulate your own fate

And as I dandle this fragile, motionable creature
His invitation is the subtle invitation of evil,
The prospect of the Immortal in a paper toy.

VI

And as I think of his perfect, immaterial existence
I am reminded of the deepest human sympathies,
While in his coyness I can make him run.
Inventiveness is his saucy mode, as he spins.

Delectable is the passionate language of the snake,
His motion has in it the stillness of all delectation,
Ancient the sinuousness, instructive the sensibility

For death is in the imagination of that tongue;
But the poison you create in entertaining him
Is swallowed up in waves of orgiastic love.

VII

Or maybe as music as an aftermath
The tantalizing poise always about to be destroyed,
Vagaries of the vigorous, quagmires of the pellucid,
Indices of the contaminations of the realities

You will perceive in fabulous sounds this serpent.
He is present to your skull as in a mirror,
It is the music of your blood you see in his visage.

Delightful is the persuasion of this destroyer,
He is the charming absolute of crumbling fingers,
The dancer, the actor fabulous in the abstract.

VIII

Self-knowledge is the tempter on the stick,
Fair is his eye, ineradicable his logic,
Unquestionable his intention to destroy,
Perdurable his ability to maintain!

For you have coiled him up upon himself,
You have looked him in his evil eye,
Adroit, you spurned him into fantasy,

And you are the pattern of his prophecy!
You are the structure of his massive eye,
And you increase his inviolability.

BURR OAKS

THE JUNGLE

It is in the jungle one finds the secret springs.
Spiny nature there, in thickets uncontestable
Coils her secrets and makes pure her spiritings.
From the opulence of the burdock the heart sings.
From the mystery of the grapevines and the plum trees
The soul will struggle to be free, inextricably
Bent in the human gaze and warmth of blood
To the folded fragrance of the ground beneath.
In the freedom of their fastness squirrels play.
It is odd, at his unaccustomed level, to look a rabbit in
The eye, beneath a tree: who is the hypnotizer, you, or he?
The fear of the thicket is what you see.
Fear will keep you from the remotest dens
Of the snakes, but will lead you to find them.
The birds by favouring denseness are contrived
To viewless melody twice their size.
The skin stings and smarts with all surprise.
The jungle, vigorous, various, interminably
Changeless, is never grown to, a foreign fastness.
You cannot see up through to the latticed sky.
To its secret strength you return again,
A mysterious place unspoiled by man,
In which you can be endlessly listening
To wildness in the weeds, the intoxicants of things.

THE ATTIC

The attic and the cedar closet—nostalgia!
Who has not dreamed evasive dreams
In the strange disorder of the attic trunks and boxes,
Fantastic clothes, broken dolls, grandfather's canes?
And in that world of broken strangeness
Had glimpses of the profound aberration of the world?
And crept in among mother's old rustling dresses
Overpowered by the cedar ether from the cedar trees?
While from the high tight windows, surveying the vast
World of trees, grove, river and pasture
It was a citadel never the world would cure,
Nor all its years of struggle, philosophy, or knowledge
Divest of a delirium of strange charm.
A special place is the attic where one goes
For special purposes, where anarchy reigns,
Oddness is the prince of disinheritance,
Angularity expresses the confusion of the senses,
Time is suicidally overcome and diverted,
Chaos is come without pain of reality.
O who has not loved hours among attic things?
Who has not grieved to return to the world downstairs?

THE GROVE

The grove is a stately place, calm and free,
Wherein the immane sky looks vastly down
On health and ease: the trees measure world's
Harmony. I have sat with my back against them
All a summer's afternoon, and shucked the big acorns.
The lawn is smooth and pleased; each grass blade
Beneath the protection of the strong burr oaks.
This is a world should never end.
It is a world controlled by men,
The trees placed for their healthful ease,
Which give the shade of happy years.
Security comes from the massive trunks and branches,

A lightness of heart from the luxuriance of the leaves.
Here the play of the imagination without curb
Combines sky and earth in knitted sanctity.
Abated would seem to be nature's processes
That the soul sing only as of holiday.
Death is nowhere to be seen in the grove,
Time has not sullied the delicate encryption
Of a forked hoof in the ground, Pan's love?
While here no sinister storms or fears
Trouble the mind with sin or tears.
O the grove, so large, so hale, so comfortable
Holds a deep music pure and noble.
Shall we not sit here always in pleasure and ease,
Beneath the rugged strength of the burr oak trees?

THE ORCHARD

It is in the orchard where a symbol flourishes.
An orchard of apple trees, mountainous white and pink in
    Spring!
Where mothers hold their small children up in the branches,
Little knowing the death time has in store for them.
I see the Maypole still, the gay dresses and the picnics,
The joyous rounds of innocent play and laughter.
How far away that vision of unparalleled splendour!
Unattainable the glowing of effulgence of the balmy scene!
Yet though all turn to worm and dust, to death and ashes,
The springtide of the rich orchard returns anew,
The fullness of the blossoms, a trust and a returning,
Fullness and solace of man against his loss,
O the glory of the orchard trees in their full bloom
And the memory of their scent keen as the skin of love,
The apple blossoms in full bloom before the apple comes.
Let it be kept as if somehow eternal and true,
Let it be believed when belief can be believed,
Against the worm in the flower, the cancer in the lung.
Let the mind hold its delicate balance of belief,

The glory come again in the memory:
Time ruined the perfection with unattainability,
Underground are the loved blossoms, the lovely ones.

THE BARN

In the barn you know the large soft cows
And they are your own best friends, even.
Their mild eyes are meant to teach constancy
To your impetuous and erratic restlessness.
If acceptance after the storms of decades
Is the ounce of wisdom in electrified flesh,
Philosophy can end in the eye of a cow.
The eye of a cow is a picture of sufferance.
The whole attitude of the animal is gentle.
It has learned the lesson of the pacifist
And by not fighting back has not been destroyed:
Even has rewarded its captor with bright drink,
A kindness never learned spiritually by man.
Deep in the obscure brain of this mild beast
Is knowledge of the opulence of mother earth.

Upon the skies of loftiest imagination man can gaze,
In no wise control the purpose and movement of the stars.
Let him then in the confines of human pastures
And upon the heart of man graze with some study
Learning from the cow the nourishment of those acres,
Supported by the supporting love of fellows.
I tossed old Daisy an extra toss of hay
When I thought of these things that early day.
Twenty years later I have thought it over
And still admire an attitude of clover.

THE PASTURE

The pasture is a field of changing vision.
Seen from a height, it stretches to distant woods.
On either side the Cedar River flows
From Baudler's swimming hole to the old sand bank.

The distant river is enchantment still,
Its least and living turn and intimacy.
In Spring it is adventure's steadfastness,
With bogs and flags; with eggs and hatched birds found;
With gophers snared in strings in deftest summer;
With list of seeds of the milkweed pod in drifting fall;
With banks of dangerous winter snowstorm snowfall.
An estate not incomprehensible, but mapped
With every zest between the barbed wire fences.
Year in, year out, the pasture is to lie in
And gaze at intellect within the sky,
And gaze through intellect to spirituality.
It is the same earth at death you will lie in,
Old earth pressing up against your new thigh,
Ancient grasses at the corner of your eye,
Air that incorporates your flesh in sky
While quite beyond the reach of men
Everywhere are treasures of profundity—
That forty years will pass away, or more,
With nothing learned in action not sensed in the earth,
In silent acts of adoration,
Prone in the pasture grass.

THE CEMETERY

To sit by graves of freshest green, and watch
The robins pass, beneath ephermerid clouds,
For long hours silent, that is the breathless ease
Of searching a high mystery,
And to be quite out of mind with ecstasy
Realizing the shortness of your stay,
The voluptuous benefit of rosy day,
And humbly to accept primordial constancy,
Your death, total decay,
Obliteration to so serene a sky
As touches now your cheek, and now your eye,
With such a fine delay
That it makes you say,

I,
Beyond the greenness of my youth and ease,
Lying by a grave under the trees
Fully conscious of my state,
Aware of the whole fate,
Accept the blue tenure of the sky,
Time's vanishing point in me,
Then young, or old,
When world's no more to see,
Let worms have deep their play
As men had their deep strife,
Let the day be other
And let

Come what may.

DAM NECK, VIRGINIA

Anti-aircraft seen from a certain distance
On a steely blue night say a mile away
Flowers on the air absolutely dream-like,
The vision has no relation to the reality.

The floating balls of light are tossed easily
And float out into space without a care,
They the sailors of the gentlest parabolas
In a companionship and with a kind of stare.

They are a controlled kind of falling stars,
But not falling, rising and floating and going out,
Teaming together in efflorescent spectacle
Seemingly better than nature's: man is on the lookout.

The men are firing tracers, practising at night.
Each specialist himself precision's instrument,
These expert prestidigitators press the luminence
In knowledge of and ignorance of their doing.

They do not know the dream-like vision ascending
In me, one mile away: they had not thought of that.
Huddled in darkness behind their bright projectors
They are the scientists of the skill to kill.

As this sight and show is gentle and false,
The truth of guns is fierce that aims at death.
Of war in the animal sinews let us speak not,
But of the beautiful disrelation of the spiritual.

THE FURY OF AERIAL BOMBARDMENT

You would think the fury of aerial bombardment
Would rouse God to relent; the infinite spaces
Are still silent. He looks on shock-pried faces.
History, even, does not know what is meant.

You would feel that after so many centuries
God would give man to repent; yet he can kill
As Cain could, but with multitudinous will,
No farther advanced than in his ancient furies.

Was man made stupid to see his own stupidity?
Is God by definition indifferent, beyond us all?
Is the eternal truth man's fighting soul
Wherein the Beast ravens in its own avidity?

Of Van Wettering I speak, and Averill,
Names on a list, whose faces I do not recall
But they are gone to early death, who late in school
Distinguished the belt feed lever from the belt holding pawl.

AN AIRMAN CONSIDERS HIS POWER

I was in the days of peace
Of warlike tissues made,
And that was a strange man
Fighting in a shade.

All things by opposites go,
Truth has there his lease,
Peace has come with war,
War that came with peace.

It is events that make us peer.
But you will veer some way, no doubt,
In balance by your natural aptitude,
My proposition quite thrown out.

Yet I know in the bomb's release
The truth I felt of the will,
From destruction is peace,
In peace the will to kill.

AT THE END OF WAR

God, awful and powerful beyond the sky's acre,
Lord God who looks down upon fighting men,
Who sees the bloody folly of them and their wickedness,
The Admiral with his cap and the private with his can,
The aviator shrouded in goggles and the marine bayoneted,
The men in the bureaus mating rubber stamps, carbonizing
    history,
The poets raving in their uncontestable ravening
The war poets who were ever at war
Before the wars, and after the wars,

Forgive them, that all they do is fight
In blindness and fury,

Forgive them for passion, and animality
Like a cloud vitiating their members,
Forgive mankind for its abominable stupidity
And fury of action,
And men for the misuse of their intelligence,
And for the intelligence that must ever be misused,
For the impenetrable fierceness you have put into them.

And may their guts justly come to rot,
The Admiral, the private
The aviator and the bayoneted or garrotted,
In bureaus the silhouetted
In pages the gavotted
Or the hooded eagles

May they all be established in the grace of time
Far away from the scenes of their crime,
Their bones awash on Himavat
Or their flesh blown apart in the air,
Translated from the terrible mummery of humanity
Into the translucence of Thy care
And done with passion, ego gone,
Self-deceiving deceased,
The unforgivability
Unpretendable,
Protect him in the pain of his erroneous breath,
Bring him to account,
Bring him to judgment,
Let the show be over and the yelling quieted.

Then, Lord God Almighty,
Visit upon mankind Thy terrible wrath
Merciless in the great eye of Thy mercy,
Altogether annihilate him, in hell,
Or in the dust of eternal nothingness,
But be Thou his condemner,

Take his life who gave it to him,
Destroy him,
Bring to an end imperfection,
And to an end his eternal frivolity.

(His passions shall be minimized
And his wars seem a dark blot
And he shall suffer plenitude
With the resurgence of the grain)

For they cannot think straight, or remember what they said,
Cannot keep their word,
Or realize how soon they will be dead,
Nor distinguish between verities
Who lust over presences,
Nor be faithful
Who are wrathful,
Nor escape animal passion
With cross-bow, slit-trench, Napalm bomb, atom bomb.

For shadows punish them when they are bright
And pride ruins them when they are strong
And time is the just instrument
That sweeps the seaman and the Admiral along,
May they never escape Thy wrath,

May they be woken up, blasted late,
Or reduced to eternal silences,
But may, pray God Almighty,
Man the cocky fighter,
The stupid and the self-destroyer,
The selfish and the vain,
The harmful cheater
The malicious debater,
The incurable self-lover
And the boring warrior,
The simpleton killer

May he learn humility
May humility be insinuated
More subtly than evil
Which brought ruin rocking him

May he turn the other cheek
Not bitten yet by worm of hate
May joy be his abandonment
In Thy forgiveness early and late

And may he learn not to fight
And never to kill, but love,
Never, never to kill, but love,
May he see Thy holy light.

A CEREMONY BY THE SEA

Unbelievable as an antique ritual,
With touch of Salamis and Marathon,
Through what visions of rebirth and death
At Atlantic's blue, hot, and sparkling edge—

War's head is up, war's bloody head,—
The thirtieth of May beats through America
With here the band, the boardwalk, and the speeches
Masking with blaze the parted beach crowds.

The traffic still restless, the loudspeaker proud,
Young couples in swimsuits strolling hand in hand
Beyond the crowd, self-interested, not attending.
But comes a hush like shimmer of summer over all,

With solemn tones the names are called, John Pettingill,
Roger Ashcroft, Timothy O'Shaughnessy, Patrick
O'Shaughnessy, Olaf Erickson, Alan Hieronymus,
William Henry Cabe, Neil Campbell, Victor Giampetruzzi. . . .

Long pause after each name, as mother or father,
Grandfather or grandmother, uncle, or brother
Awkwardly walks through the entangling sand
With sheaf or wreath of flowers to the flower-waggon,

Burgeoning with brightness beneath the bandstand.
The newly dead! The young, the dead far away,
In the strange young reality of their war deaths
Too young for this austere memorial.

The last name is called, the last flower is funded,
As people stand in the daze of the actual,
Then eight young men of the Army and Navy,
Almost naked, strong swimmers of the tides of life,

Their muscles blending in among the flowers,
Take up, four each, the flower-twined ropes.
Like a mad disturbance, hafts to the hilt of earth air,
Eighty Corsairs plummet out of the sky

From high inland down over the bandstand,
Four abreast, flash out over the sand
Over the ocean, and up easily and turn.
The deafening noise and closeness is a spell.

Then the flower waggon by the stalwart ones slowly
Through the crowd begins to go to the sea,
And as it draws, the ocean opens up its heart
To the heavy hearts of mourners by the sea's edge.

These, all kinds and conditions of men,
Thereafter follow across that bright, that transient course
As they would pay their tribute to the waves,
To the justice unaccountable of final things,

Followers of flowers; the gorgeous waggon-coffin
Drawing the blood out of the crowd, slow-passing it.

And lastly I saw an aged Italian couple
Too old to stumble through that catching sand,

The backs of their bodies bent like man himself,
Ancient as Marathon or as Salamis,
New and ancient as is America,
Diminish with laborious march toward the water.

From the platform funereal music played
In dirge that broke upon purged air
While now in mid distance, far by the shore, the new fated
Stood; then the powerful, the graceful youths

Like gods who would waft to far horizons
Drew the waggon in, which then a boat of flowers
They swam with, dim swan on the searoad.
A waiting gunboat fired a last salute.

WORLD WAR

Flutesong willow winding weather,
Tomorrow lovely undulant today,
Glorious bird glide in forest glade,
In meadow golden lissom girl dance,

Tremble air with never yesterday,
Grassy twirling boyfoot triumphing,
Budding bough drops lovely lording,
Pearling cuckoobrook cool ecstasy,

Woven from lucid sunny nest,
World of mellow willow mist,
Now forever pleasures piping,
Honey supple body wonderful:

Strike down, batter! shatter! splinter!
Destroy! fracture! cripple! butcher!

Knock! beat! whack! cuff !
Ruin! gash! smash! blast!

Baby Red Breasted Chained Nippled,
Pavement Clattering People Crippled,
Youth Courageous Finger Felled,
Nutty Manhood Maggot Shelled,

Buzzards Smiling Char the Sky,
Pain Caressing Bites the Eye,
Grass Has Adders Time Vipers,
The Heart Burns its Lifted Ladders,

Howls the Whirlwind Over the World,
Tempests Quaking Shake the World,
The Earthquake Opens Abrupt the World,
Cold Dreadful Mass Destruction.

BROTHERHOOD OF MEN

I

Caught there, then, on the Rock. At Corregidor
Caged with the enemy, sleepless in inexactness
The mind wavering with wishes of decisions,
Fear frequented the bowels in their fury,
Actor I was, avid of answers, egress,
Hungry there in a hovel, had to steal food,
Harry inventiveness, deft of devices.
Had to creep by the inching invaders
Lifting my life in my hands, lily-livered,
With motive of man's ancient menace
Dared danger, gave to a dying friend.
Often eased the enemy, agile of acquiescence,
Stole drugs from the cupboard, mates to maintain.
Saw tortures, and tall tales expedited,
Urged the underground, close escapes lived with.

Lined up with the blistered, the intrepid,
Passive at bayonet point in the belly.
Flinched not, was cast down with the feckless,
Pretender on purpose, mastering machination.
Thought of home and youth as sickness prevailed,
The shining sitters at stately banquets,
Lofty in lapsing dreams as dooms eventuated,
Blind in trust, wondered forever of survival,
Cupped heads in my hands at baleful midnights,
Waxed wary only companions to save,
Thought nothing of self, nightmares of dyings.
Bones softened by black malnutrition,
Bloated were heads and legs ulcerous,
Never gave in to the enemy, found fierce
Store in physical love for dying brothers,
Wondered on wakening how I survived oft
With strength to stall the staggering.
Played opossum to the enemy's piercing examinations,
Smile saintly, collaborated clearly,
Never gave in to the enemy, death was common.
Came ruthless purgation of the pale living,
Undreamed of excruciations, excrement of prison,
Diseased men doled on the death march of Bataan.
Fatigue was early, fierce was the fury,
Blazed by bayonets side-fellows fell,
Stumbled stalwarts, cursed many,
Crazy became some in the heat of hatred,
Infernal the features of famished men,
The enemy edged us in, egged us on.
Violence of life in the vividness of madness,
Kept by miracle of man's sufferance
Some semblance of order, reduction to essence,
First blanch of weakness was first broached.
Down to the doom of man's mastery of self,
Fate was fearful, fate dandled the balance.
Strong those seemed who cracked first, cried,
Crumbled in clairvoyance to butt-stroke of rifle.

Miles were nightmares, water was absent,
Mercy was nowhere, but stood the gaff,
Wonder of wonders, never knew how.
Hardest was not to move muscle
When friends were violated; vicious visitations
And violent strokes ebbed the naked strength.
Eyes should not see such sanguine spectacle,
Death not dare such darting depravity.
Retaliators were whipped, naked were whacked,
Were cuffed quailing, pierced with pikes.
Soonest were doomed those who cared aught,
Care for others in the carking tribulation.
Kept as if eventless, played against the enemy,
Looked not sidewise, helped not the helpless;
Cunning of the canny bred continuance.
Survived the death march; men are makeshift,
Weary of wishes to kill every one, weary
The enemy of watching comedy of twitches.
Favored by nature were the wan survivors.

II

Came lassitude and despair of the mind,
Mounted months of weakness, filth of fellows,
Pain of realization, poison of water.
Death was daisy, the mind followed fictions,
Spirit was luckless, lack of news told.
Emaciation was the flag of emancipation
Death waved at the doors always closed.
And then was transferred to Cabanatuan.
Sweaty and reckless, but holding to truth
Thought of turn of the nightmare tide,
Was able to steal in the stalking midnights,
Gave grace of grain to lads groaning,
Maggots were many, relished with rice,
Bitterness of necessity, better nourishment.
And thought of home hale in the States,

Impossible presentiments of feverish dreams,
Kept with all incommunicable dreams,
Vast vistas of times impossible.
Heard months later of fates foregone,
Heard the curt shots of ones caught.
Mentally deranged were many, no mail came.
Enemy was endless, workers were listless,
The down were soon out, the dead were left days.
Turned at night to clasp your friend dying,
Shielded him; woke clasping a death's head.

Monsters were moving; sickness established,
Endurance was endless, never was mercy.
Were mild, many, with infection, lips were livid,
Heads swelled like cabbages before the soft death-rattle.
Came day of dispatch, was chosen with others,
Knew nothing, but was sailing by ship.
Rumors ran, shipment of prisoners to Japan.
The heart lifted somewhat, but dazed was desire,
Man suffering much cannot believe early.
Little knew those of the perils impending.
Haggard by hundreds, thrown into the hold,
Heat staggered the stifling; time wore on
As we wavered in the knowledge of the water,
Packed like pigs, herded in jeopardy,
Mass and mess of mankind standing up,
Heat harried us, parched were people,
Panic mounted in pandemonium,
Merciless was the murderous sun, murderous
The enemy with machine-guns at the hold's top.
Days unendurable, the dead fallen were trampled.
In desperation, doughty, doomed countrymen
Climbed on others, climbed the hot iron ladder,
At top were shot down, fell on their fellows.
So close were packed all was a clenching,
Madness was manifest, infernal the struggle,
Urine was drunk by many, rampant was chaos,

Came wild men at each other, held off attackers,
Some slit the throats of the dead,
Drank the blood outright, howled wailing,
Slit the wrists of the living, others
With knives, or with fangs ravenous,
I saw them drinking the blood of victims,
Hell I was in, this was immitigible Hell.
Endless hours fought off famished, crazed attackers.
Savage the senses in dreams of delirium,
Never knew how I surrendered not,
Saw clearly my mother in the midst of terror:
'Persevere. Persevere. Persevere. Persevere.'
Faith beyond reason, wrecked beyond words,
I dogged clung, daft before danger,
Dazed by disaster, damned in the hold.
And hung on God knows how, saved by my nature.
Some peculiarity of nature saved me, hundreds perished.
Would never have won, but our own planes
Came hungry, hunting enemy ships.
Knew not our own men were therein.
Bomb blast in the hold next killed all there,
We were as if raised out of water.
Clambered to deck tottering and reckless,
Gone were victors, machine-guns vacated,
Like bales of lumber dead were stacked on the decks.
Caught handfulls of sugar from wrenched sacks,
Stuffed mouth, pockets; jumped over, water revived me,
Saw the shore and swam to save me.
Some seventy swam, of all holds' hundreds.
Crept to a clearing, fraternity of survivors.
The enemy entered the clearing, commanded us.
A marine's leg was shattered, smoked calmly
One cigarette discovered; clasp-knife severed his leg.
(Heroes were many). He died gamely.

A summed spectacle in the sultry sunlight
Our sapped band was, after hellish onslaught.

We stiffened our resources, bonds were breakless.
Speechless were the depths of humanity,
We were the living, no time for understanding.
Again the enemy did not kill us,
Ammunition doubtless sparse on this island.
But took our shoes, practically fatal,
None could hope to escape without them
And live long in the barbed jungles;
Soon would be forfeit to tortures of thorns.
Weakness persisted, rest was undertaken,
Watery rice with maggots always our ration.
Now numbness of spirit prevailed, what more
Could happen? How long could man endure
Unknowable fate? A ship came to take us,
This time without torture we got to Japan.

III

My mind was heavy and my luck was dark,
From chaos I came, to greater chaos went,
Stubborn of spirit, subtle suborner,
Four years in prisons and vile execrations
I endured, and worked to save my companions.
Fame I never had at home among fellows, was never
Known as the bright who knew where he was going,
Suffered severely from cul-de-sac sensations,
But I had animal cunning, abundantly,
Reserves I had that were boundless, idiot cunning.
I did better, being instinct with the basic,
Undeluded of delusions, had knowledge of nature,
Queerness quested my queerness, madness not maimed me,
Basilisk my eye baleful to baneful totality,
Delicate my decisions due to no dearth of nature,
But to strangeness I gave ever strange nurture.
The normal, those with notions pre-conceived,
Died early, or died late, but always died.
To test them so far nature never intended,

It took madness to fight madness, savage simplicity
To slay savagery, sleights of mind for savagery,
Tricks of balance and tenacious trickery.
Fared lax like one in the forfeited herd,
Yet undefinable nature used my example yearly.
I, weak, had strength beyond the strongest,
Wounds I wore out, amazing to myself.
Held, helpless, to the help of Heaven,
Expected death at every ditch,
Only hope was the help of helping others.
I observed those survived best were feminine,
Fared on duplicity, dark deceivers;
Most male were most manhandled, maimed.
Womanish men were puckish, pictures of Psyche.
The enemy sensed their lack of enmity,
Entered they had into subtle sympathy,
Fragile balances saved them savage bayonets.

Winter came in heathenish hell-holes,
Water was wicked, cold was killing,
And never knew how the war was going.
Slept with diseased brothers to keep off the cold.
Low was language, life left us slowly,
Some sank into states of sodden trance,
Longing to lie in cold bed and to die,
Weary of spirit, would not wake, will-less.
Warmed the waning, months at a time,
In coughing nights of incurable cold,
Tempted to bend to the bed of death beckoning
Beyond disease's ravage, blessing of rest endless.
Peculiarity of nature nerved me nevertheless,
Locked in bunks befouled with bloating brothers,
I forced fellows to follow me, fall on the floor,
Get out of beds where they were breathing only,
And stand in dark days against compulsion of death.
Broken bones were left to brokenness,
Festering flesh was guest to feebleness.

Work was a wisdom and kept the mind alive,
Worked for nine months sewing buttonholes.
Long had forgotten of justice, but never
Of justicing; never gave in to the enemy,
Never surrendered, the enemy was myself.
Monthly made him think I loved him,
In some strange manner I loved my captor,
I belonged to mankind, all were responsible.
In his place would I not have done the same?
My bayonet would not have pierced the plucky,
Hatred would halt somewhere, heavy of heart.
We were only soldiers doing our duty,
All imprisoned in primal curses,
All commanded by evil in man's nature.
Learned in the longest years of my life,
In depths of dejection, disjunctive thralls,
Grains of good in the coldest killers,
Quest of kindness of the unbelievers,
Qualms of conscience, costliness of guilt.
Enormous it was, and frail were all,
Victim and victor alike vacillated.
Error was everywhere, ever flagellant,
Freakish the force of death or survival.
Each event and trial was unique,
No law held for all operations.
In gruelling slowness of gravid time
Never saw women, forgotten were children.

IV

Rumors of liberation. We could not believe it.
Liberation came. Planes came over parachuting
Packages. One plummeted through a sky-light,
Broke one of our legs. Greedy as children,
We ate chocolate until we were sick,
Suspect bellies could not stand it.
It would take us years to get well,

Our bones soft and easily breakable.
My hand broke, opening the door of a car.

Rings I have, watches, tokens, a dog tag
To take back to the land of the living,
From the dead to deliver to fathers or sisters,
Cherished possessions of my luckless companions
Lost in four years of rooted abuse.
O to forget, forget the fever and famine,
The fierceness of visions, the faith beyond reason,
To forget man's lot in the folly of man.
And swear never to kill a living being,
To live for love, the lost country of man's longing.

And yet I know (a knowledge unspeakable)
That we were at our peak when in the depths,
Lived close to life when cuffed by death,
Had visions of brotherhood when we were broken,
Learned compassion beyond the curse of passion,
And never in after years those left to live
Would treat with truth as in those savage times,
And sometimes wish that they had died
As did those many crying in their arms.

INDIAN PIPE

Searching once I found a flower
    By a sluggish stream.
Waxy white, a stealthy tower
    To an Indian's dream.
    This its life supreme.

Blood red winds the sallow creek
    Draining as it flows.
Left the flower all white and sleek,
    Fainting in repose.
    Gentler than a rose.

Red man's pipe is now a ghost
　　Whispering to beware.
Hinting of the savage host
　　Once that travelled there.
　　Perfume frail as air.

'GO TO THE SHINE THAT'S ON A TREE'

Go to the shine that's on a tree
When dawn has laved with liquid light
With luminous light the nighted tree
And take that glory without fright.

Go to the song that's in a bird
When he has seen the glistening tree,
That glorious tree the bird has heard
Give praise for its felicity.

Then go to the earth and touch it keen,
Be tree and bird, be wide aware
Be wild aware of light unseen,
And unheard song along the air.

'SOMETIMES THE LONGING FOR DEATH'

Sometimes the longing for death,
Imaginative death, comes hard.
Not the unimaginable, last gasp,
But lift to supernatural love.

And sometimes death is achieved
In the visionary place of the mind,
New life, painful no more,
Where hope does not need to deter us.

106

Is it for this, most personal, most secure
Life turns, in goodness, and in evil
To tear our wits out of our wills
To its built, tremendous godhead?

AT NIGHT

In the dust are my father's beautiful hands,
In the dust are my mother's eyes.
Here by the shore of the ocean standing,
Watching: still I do not understand.

Love flows over me, around me,
Here at night by the sea, by the sovereign sea.

Gone is that bone-hoard of strength;
Gone her gentle motion laughing, walking.

Is it not strange that disease and death
Should rest, by the undulant sea?

And I stare, rich with gifts, alone,

Feeling from the sea those terrene presences,
My father's hands, my mother's eyes.

A LOVE POEM

I am the lightness that I know
And I am the terror that I seem;
The taint, the question, and the dream.
Perhaps you are also.

When the wind blows, and trees shake,
I think there is some comfort in them;
Else, in the silence of contemplation,
All harry me and my mind take.

Therefore, what I am interested in
Is the anatomy of time;
Whether there is anything substantial
In the world and flesh we are in.

But Love told me the answer,
In the moist garden, when the worms were out;
Love evaded the painful question,
When my eyes were a green dancer.

GOD AND MAN

My grandmother said I was an atheist
God said I was a man.
My father took me by the hand,
My mother vanished in a mist.

In the rich stores of the ether
The future was seen as the past,
Flesh was aerial prescience,
And the devil was seen last.

In time you have no grandmother
For ancient earth recedes.
Your father and your mother go,
But God says, you are a man.

Between budding leaf and blue sky
Angels of mercy were spreading,
Like bees around the cider-press
Diffusing this blood with murmurousness.

The angels were the archetypes.
Would go away while devils overcame
Time, and chased the crooked years,
You still lusting after evil.

Every one a father and a mother has
And every one more ancient staffs,
Yet all lose even loss itself
When God says, you are Man.

For man precedes his knowledge
Aroused within his variant myth,
A stalwart, fiery with animus
Whose death is only another dream.

And God has the deep justice,
And God has the last laugh.
To be God God needs man
As man needs God to be man.

THE HORSE CHESTNUT TREE

Boys in sporadic but tenacious droves
Come with sticks, as certainly as Autumn,
To assault the great horse chestnut tree.

There is a law governs their lawlessness.
Desire is in them for a shining amulet
And the best are those that are highest up.

They will not pick them easily from the ground.
With shrill arms they fling to the higher branches,
To hurry the work of nature for their pleasure.

I have seen them trooping down the street
Their pockets stuffed with chestnuts shucked, unshucked.
It is only evening keeps them from their wish.

Sometimes I run out in a kind of rage
To chase the boys away: I catch an arm,
Maybe, and laugh to think of being the lawgiver.

I was once such a young sprout myself
And fingered in my pocket the prize and trophy.
But still I moralize upon the day

And see that we, outlaws on God's property,
Fling out imagination beyond the skies,
Wishing a tangible good from the unknown.

And likewise death will drive us from the scene
With the great flowering world unbroken yet,
Which we held in idea, a little handful.

THE TOBACCONIST OF EIGHTH STREET

I saw a querulous old man, the tobacconist of Eighth Street
Scales he had, and he would mix tobacco with his hands
And pour the fragrance in a paper bag.
You walked out selfishly upon the city.

Some ten years I watched him. Fields of Eire
Or of Arabia were in his voice. He strove to please.
The weights of age, of fear were in his eyes,
And on his neck time's cutting edge.

One year I crossed his door. Time had crossed before.
Collapse had come upon him, the collapse of affairs.
He was sick with revolution,
Crepitant with revelation.

And I went howling into the crooked streets,
Smashed with recognition: for him I flayed the air,
For him cried out, and sent a useless prayer
To the disjointed stones that were his only name:

Such insight is one's own death rattling past.

SEALS, TERNS, TIME

The seals at play off Western Isle
In the loose flowing of the summer tide
And burden of our strange estate—

Resting on the oar and lolling on the sea,
I saw their curious images,
Hypnotic, sympathetic eyes

As the deep elapses of the soul.
O ancient blood, O blurred kind forms
That rise and peer from elemental water:

I loll upon the oar, I think upon the day,
Drawn by strong, by the animal soft bonds
Back to a dim pre-history;

While off the point of Jagged Light
In hundreds, gracefully, the fork-tailed terns
Draw swift esprits across the sky.

Their aspirations dip in mine,
The quick order of their changing spirit,
More freedom than the eye can see.

Resting lightly on the oarlocks,
Pondering, and balanced on the sea,
A gauze and spindrift of the world,

I am in compulsion hid and thwarted,
Pulled back in the mammal water,
Enticed to the release of the sky.

THE CANCER CELLS

Today I saw a picture of the cancer cells,
Sinister shapes with menacing attitudes.
They had outgrown their test-tube and advanced,
Sinister shapes with menacing attitudes,
Into a world beyond, a virulent laughing gang.
They looked like art itself, like the artist's mind,
Powerful shaker, and the taker of new forms.
Some are revulsed to see these spiky shapes;
It is the world of the future too come to.
Nothing could be more vivid than their language,
Lethal, sparkling and irregular stars,
The murderous design of the universe,
The hectic dance of the passionate cancer cells.
O just phenomena to the calculating eye,
Originals of imagination. I flew
With them in a piled exuberance of time,
My own malignance in their racy, beautiful gestures
Quick and lean: and in their riot too
I saw the stance of the artist's make,
The fixed form in the massive fluxion.

I think Leonardo would have in his disinterest
Enjoyed them precisely with a sharp pencil.

FORMS OF THE HUMAN

I wanted to be more human
For I felt I thought too much
And for all the thinking I did—
More rabbits in the same hutch.

And how to be more human, I said?
I will tell you the way, I said.
I know how to do it, I said.
But what I said was not what I did.

I took an old garden hoe
And dug the earth, and planted there,
Not forgetting the compost too,
Three small beans that one might grow.

Three grew tall, but one was wild
So I cut off the other two,
And now I have a wild bean flower
The sweetest that ever grew.

OEDIPUS

Oedipus should have found exit from his dilemma:
Blinding of the eyes does not improve the insight.
How dark and magnificent were the Greeks! How
Their melancholy rings down the years to this day,

Rings in my ears the wail of the man going blind,
As universal blood streams through the firmament.
Not to unravel the doom but to accept it
Is even what I did when first I saw him bleeding.

Oedipus should have had a will to escape it.
He should have been able to outspan the plot.
Is there no perception to pierce through Fate?
Is nothing bright in man to dispel Destiny?

In barking and brittle light, in the antique air
I see him in the universal drench,
Fate that is stronger than the soul of man,
Destiny that rides us, whirlwind-like.

FRAGMENT OF NEW YORK, 1929

Four-thirty in the morning,
The coldest morning in late October.
Mechanical,

I stepped into clothes sprawled at hand before.
Effigies of tenuous waking dreams
Walked with me down the stairs,
Hardly realizations,—subtle separate sensations.
Chill air, and congealed skin,
So wake the mouth and hair.
I walked to work, Times Square
And all buildings little comprehended.
Unreality again, all this steel,
Height, lights: my feet alone know.
The usual stragglers pass. Strugglers?
Or only I? Or not I, they? Is
Nothing palatable but figs and coffee?
Then skirted along, briskly
Forty Second Street, to the west.
Head down, wind like a wire whip
Trying from neck to chest. Fingers
My collar closing are like a python's mouth.
My thighs feel good,
Action always feels good.
Except contemplation's,
When feet run in the brain.
Tenth Avenue, newspapers asleep in humble blankets.
Why look back? or even forward?
Feet can tell. Ten minutes late
This morning; see, night opens
One shade to the morning. The stars merely
Are pin points; not as on the Indian Ocean,
To be waked, naked, warm,
On a hatch, when stars sigh goodnight.
Not very wary. Eleventh Avenue,
I came around the corner and there,
Above the seven box-like and lamp-lit
Shelves of the packing house,
Glared the visitant metallic moon,
Canny as a dead man's eye.
Taormina. God, floods of foam.

There's a plumber playing pipes in my guts,
It screeches. Escapeless, common feeling.
Vapour from the factory rose up
Against the cold and paling light,
But vanished lower than the moon's aura.
Why must it be thus with you, never
To find the one untormented integer.
Surprise moon, four-thirty moon,
Dead moon, so rich with our ecstasies.
My stomach's got the goose flesh: wind of thin whips.
Into Hell again. I'll one glance,
Two glance, let it go, slip the moon's pierce,
Or grow great with old ravage.

Until the din became immense and profound,
A living horror of common death,
Merciless, disjunctive, co-ordinate;
Until relativity made this equable, and
Rhythm was levelled with it, it had
No meaning but simple action, happening.
Only one in a hundred kicked;
Kicked violently, pathos-maker,
Leer-long, with every energy.
A hog became a living curve,
Drew itself convex, from the vertical
Imposition of the ankle shackle.
Everything in nature curves.
So with beauty also; but I never
Decided; or decided so many times,
I have no decision. Decisions
Are practical, they do not increase the soul.
The killer's face! He is baffled now,
Seems. Moment. He poises
The tip of the knife at the throat.
So little is life. He cannot make
The one swift entry and up-jab.
Curious copulation, death-impregnation.

How the blood flows out
Dark red in a full stream;
Or some hold theirs in, muscles so tense,
One minute, then bewildered they rush into panic,
Six or eight or maybe only four
Wild screams, or grunts that can't
Get through the blood. This one baffles:
Odd. The vibrating free leg's too quick
For the killer to catch it. Then,
With almost a solemn decision,
The life-taker jams the terrified animal
On to a side rail: three more wait, with
Din-horrific screeches. They are dispatched
With able, easy gesture,
Purposive intercourse!
When there is a pause in the chain,
His inevitable great hand swings round
(Belly to belly) the daring pig.
Still its agony increases,
So much vitality expended,
Every muscle expresses itself
Against imminent death.
A dramatic instant of pause.
Of course there is no hope. In
Some dark way I wish
Life were less terrible.
These are all people strung up,
You understand. Not one chance,
Not a single.
Nature wants no virginity:
Must wed and be dead.
Big purpose (pig purpose) is to be killed.
We hope the inevitable hand
(Splendid huge veins there, bright wet
Red to the elbow) will be stayed
From him, from us. What a magic instant,
The tip at the throat, a little jab

Into the skin unfelt, then the
Deft probe, the jugular, conception.
Thus! And the admirable God's creature
Dies powerfully screaming.
A show! A show!
Come see the big Negro stick pigs.
That one was rare, to kick against the pricks.
His enthusiasm was unwarranted,
As if he were not of his genus,
But had the life of genius.

This is a vast sadistic enterprise.
Good for us, times, to see and feel
Down into sixty men. They
Stand upon greasy benches, with sharp knives.
The endless chain is begun,
With guttural as-if-eating-bolts groaning.
It is the endless chain of their labour.
Chilled hogs appear at the end
(As in a Blake angels would
At 'per me si va tra la perduta gente').
What if they, too, were people,
Some other intelligence were using,
As are used, these carcases.
The earlier initiation,
Neophytes knowing their Achilles heel,
Their tragic flaw, heads downward to earth,
Great Mother! All dead by
Some Pig Intelligence butchering.
Guts ripped out, young women too,
Old men, after fresh bath in a hot tank
Scalding the hair off. Then the
Whole family, and cousins Kate, Bertha, Percy,
Strung up, as I said, by Achilles
His heel. Ripped down
From penis, or vagina, through navel
(Inside out) to chin or cancer of the throat.

Faces sliced off, see fantastic mirth,
Comedy of errors, tragedy of desires,
On expressions so quick-changing.
Intestines on moving tables,
The last humiliation of having
The insides examined. If foul,
Condemned. Skinned, fat cut off,
Around we go on the chain slow,
And a great knife down, once, through the spine;
The skull split open, brains given an airing,
At last. Pea-sized pituitary.
I said the chilled hogs appeared at the far end,
A Negro big as Amazon,
Amazing, cuts them down onto the belt.
A white man, next him, like a giant
Marionette, with titan thews and blows,
(Conditioned by recurring necessity)
Chops with one vast climax
Belly from loin. His body
Turns, shifts, turns, swerves,
The splendid toy,
Man full-strong and hale.
He loves his home's children.
Also alcohol. Plato's dead.
Butchers are told by their knuckles,
As poets are told by love.

On top, all storeys up, if you walk out
On the roof as if doing something,
There are the ships,
The lambent air burdened with life tunes,
Little screaming and quiet symphonies,
Even in New York. O tall beautiful,
No, not sails any more, nor Queequegs:
Yet not too sullen steel plows,
And carriage like health. The
Colour of intestines fascinates.

I saw lungs like mackerel sunsets in Haiphong;
Smooth bladders the
Shade of evening's amber rain;
Hearts as of New Mexican earth,
When wet; all very carnelian.

Why love so long the even, but almost
Deathy darkness, that wells
Into dawn over buildings:
Opaque quietude, such tonal,
One mode. Why love so strong.
Perhaps because gentleness
Looks real (like death).
And the sun, when he comes glaring,
Is pain unending. Wrap around me,
Shroud of the dawn almost:
The sun with his eye will put out mine.
Sweet universal sky, seen
Through the sesame-slit of a dirty lavatory,
Where I wash my hands of hates.

I pretend I am the young Goya,
Enjoying fallow and burnished,
There, polyglot faces. His the bigger
German's, with hands oily to match,
Of matchless golden aura, dark
Enough, as in Rembrandt, too
Solid to be eclectic, no Goya;
But smoothest tempered skin all
Health, an animal pleasure,
Let us rejoice. Undangerous he,
Regular, in simplicity, without
Pride but humble in muscles
And can say "aw right" almost American.

John, Czecho-Slovakian competent,
Bites his hand in a steel door.

Incoming truck's the gorge, bitter
Teeth, and no fault. Pitiful
O my John, like a different man
With mind bemused and maimed,
John like a boy almost whimpers,
Waits upon sympathy. Puffed
Hands under warm water tap.
Huge arms capable of
Two hundred pounds hold the hurt
Hand as it were a flower;
Or a dead something.

There is some mournful music,
Somewhere, I give myself into
The slow god. Like a man
Grasping at peace, maelstromed
In this fourmillante cite.
Until sombrely, lovingly I close
In the doleful scales my being.
It is a whole world of direct power,
Now, intangible, being
So in solitude, seeming powerless.
But while in the thick pleasure
Not any says: experience
Lives as at a sea's depth,
In its darkness, surrounded
By itself, by itself filled,
Tonal tropism. I am
My own tonal tropism. I am.
The something done by dense
Minor, no lyric, oneness,
One sound, and essence of being,
Being without the object, yet
So surely oneself in spite of it.
Accordance of sympathy,
Equilibrium of the tempo composed
Of disparate things. All sorrow,

Let it be my shroud, in folds
Of music. Known in the heart's
Well, like standing in mud.

Embowelled again. Put the insides out.
This is a vast peristalsis.
Robbers dine on expectations,
In caves of hesitancies.
Sparse gigantic flecks of what might be snow;
Walk on murder trials to work. It is so cold;
Try over the buildings to the colder moon,
That same, for warmths, fires, flames of hell.
Not there, this dated night, can keep love
Out of mind, Sicilian singing.

In the lyric buoyance of this dull café,
She came into the café, walked to the back,
Sat with her back to the front
Of the room. The ugliest woman I'd
Ever laid eyes on,
Life to the viscera. Revulsion and vision spring.
Skin like malice, that makes her eyes unbearable.
Ripeness is gall. She spoiled my meal,
And a rat ran over my feet.

I thought they were all standing
Around like dead men,
So much physical life they had,
Who laboured among red substances.
They were dead men, so I mused
Among noises, silence the heat and
Cyclopean eye of vibrations,
Activity the opposite of whatever it is.
They were dead men, with big hands
Cooler-puffed, no life in them,
Continually I mused upon,
And amused me with this

121

Runlet of consciousness,
Confounded with the actual?
No life in them, but they were of life,
See still I fret about the mind.
But one might almost say their bodies thought,
In their honest and cool cells.

I have flung myself down into this pit.
Now they are all dead; statuesque
Stand these actors, these buildings
Building no more. Death I saw,
And wormed through it. And make fragment
Of the end of a time, when seethed
So thick the life, it knew not,
In savage complexity, modernity,
The harsh omnipotence of evil.

AESTHETICS AFTER WAR

(*To A. Nykyforcza,*
Student of Sighting.

The floating reticle became your eye,
You saw flashing battle,

You returned in the death lists.)

I PROPOSITIONS

Is the rose the same after it is seen?
Is it brighter if the seer has a blighted gall bladder?

If a poet is colour blind, either by nature,
Or say by choice,
So that his deep wish to see purple
Gives him constitutionally a purple world,
As the romanticist finds excess where others

Do not find it,
As the classical scholar finds the world more classical
Than any striking steel worker knows it,
As the aesthetician finds the world
An aesthete's paradigm, to whom then
The pleasure principle is the end of all,
Do the poet, the romanticist, the classicist, the aesthetician
By their profound aberration
Discover the true reality of nature?

What is the relation of aesthetics to philosophy?
Should there be any?
Is its branch an authentic olive tree?
If you contemplate the rose, do you have to think about it?
If you achieve a beatific state
In the contemplation of the rose
Are you loving wisdom?
Is it possible to achieve abstract purity,
The ultimate knowledge of the object rose,
Without a mystical infiltration?
Should an aesthetician wish to think?
Should a thinking aesthetician want to know God?
Will God appear in the ultimate stillness of the rose?

Keats confused, confounded two centuries
By ambivalent, ambiguous
Mating of truth with beauty.
Or are these absolutes safe in unattainability?
So that the searcher, as centuries ago,
May struggle, physically, logically, semantically,
Or by purposive derangement of the senses
To find them, being relatively certain
Their abstraction will inhere and remain
Long after his bones have turned to very earth?

What has the aesthetics to do with society?
Was the Italian airman crazy

When he saw aesthetic purity
In bombs flowering like roses a mile below?
He could not see nor feel the pain of man.
Our own men testify to awe,
If not to aesthetic charm,
On seeing man's total malice over Hiroshima,
That gigantic, surrealistic, picture-mushroom
And objectification of megalomania.
A world of men who butcher men
In the arsenical best interests of several states,
The modern warring maniacal man,
Is this world of men inimical
To the postulates of the study aesthetics?

II INSTRUMENTS

There are many intricate pieces of workmanship,
Precise instruments like the Mark 18 Sight,
With a floating reticle and a fixed reticle,
The fixed being a circle of light with a cross at the centre,
The floating being eight brilliant diamond points of light
In a broken circle which enlarges and contracts
Framing the enemy wingspan, increasing
As the enemy plane comes nearer, grows larger,
Decreasing as it flies away, grows smaller,
The whole floating reticle a dream of beauty,
But accurate to a split second of gunfire,
Its gyroscopic precision solving all problems
Of boresight, the pursuit curve,
Of wind drift, range, of bullet pattern
So that as semi-automaton all the young gunner
Has to do is to frame the enemy plane
In this brilliant circle of light and blaze away.
This is one of a bewildering array of imaginations.
Radar, as another expression of ingenious invention,
In its excessive, but already dated modernity,
Only gives man what bats have used for centuries,

Whose vibrations, beyond the reach of the human ear,
Strike obstacles which echo back to warn the bat
Who instantly evades each thing that would harm,
Although he seems to us to employ erratic flight.

Warfare spurs man to electrify himself with technics
But never can the human be contraverted
And as the Mark 18 Sight is only the fastest
Eye in the fastest brain, the perfectly anticipatory one,
And as radar is only an equal intelligence
To the ancient, instinctive knowledge of the bat,
So mankind in his abrasive rigors
Constructing the mazes of his complex aircraft
Often unwittingly makes them look like huge
Beetles or other insects, and I have seen
Hundreds of Corsairs parked in the evening glow
Their wings folded back ethereal as butterflies.
If a floating reticle in an electrical sight,
If a radar screen with its surrealistic eeriness,
If an airplane poised on the ground like a butterfly
Are beautiful, is their beauty incidental?
Is it man's limitation that his mechanical creations
Perforce cannot escape from his manhood?
And that try as he will, his works are human
And never stray far from the functions of the natural?
The mystery is whether the object
Mystifies man,
Or whether the mysteriousness within man
Transubstantiates the object;
Whether the world is finally mysterious,
Or if the Deity has put a mystery in man.

III THE PULL OF MEMORY

Each argument begets its counterpart,
Only in opposites the truth is human,
Intelligible, the shoemaker Boehme said.

I recall a steamer on the Pearl River
Slipping out from teeming Canton,
Hong Kong outward bound through swarming sampans,
The glow of the East, the intense hot day,
The swan sweep of the boat on swan-swept water,
The lull of the hours in the yellow afternoon,
I remember walking the deck
Watching the ritual of the opium eaters
Through glassless windows in the inner sanctum,
Sensing their subtle gestures and serene manners
Through the prolonged trance of the opium haze;
Then looking out, on the banks of the Pearl River
Ancient of days, and of centuries,
There stood a tall pagoda old as the memories of China.
So Buddha seemed in the soft air to dwell,
Incomparably indwelling, selfless and whole,
Without action, away from world's suffering.
And I was stolen in a trance
Of the pagoda like a jewel in ancient, shimmering air,
And of the mild-eyed Chinese recumbent after smoking,
Mysterious inducements of the suffusing scene.

In the East contemplation is a self-annihilation,
In the West it never escapes intrusive action.

IV REALITY

The tail, the waist, the nose turret or the ball gunner,
Using the sight with floating reticle
Has only, if his work is properly done,
Has only to press the trigger. The enemy explodes in the air.
It is all so fast sometimes
Neither pilot nor gunner can see the result,
Until, far up and away
Banking over, high above the blue ocean
They glimpse far below pieces of plane
Drift idly, suspended in the air.

But reality is there.
Death is the reality in this case,
Love is the reality of St. Theresa,
Identification is the reality of Boehme,
For Blake innocence and experience
Were indistinguishable in mystical affinity,
Our enemy pilot was dead by a death-dealing pounce
Of superior machinery and superior manoeuvre and aim—
But what Chance was there!—
For in battle, as if man were made of adrenalin
There is no time for fear or fault, for faith or fame
But pilots say it is all like a football game.
Back on the carrier their hearts may pound,
After the event, when significance comes,
But up in the air intense and free
Controlled and able,
Perfectly secure,
What is the end of a man
You never saw before,
You never see?

It is the end of this man whose life yours never touched,
Of whose existence you never knew,
Young man, young man,
Whose floating corpse you will never view,
Whose friends you also slew,
It is the end of this man,
Or let us say it is the end of Man
Christ shed his blood for,
He shed His blood for you.

He knew you, savage trickster,
Your accomplished guile,
The total ignorance of your intelligent blasphemy,
The evil that is ineradicable
He died to show.

He would redeem the enemy airman's blood-muffled scream,
He would redeem the pride of your indifferent victory.

Is there any doubt that Christ was the most aesthetic man?
As aetheticism is a part of philosophy,
Philosophy a part of life,
Life action, for even the Nirvana-seeker still breathes,
And Stylites pulls up food in a basket,
So Christ contemplated the ultimate origin,
But originated the ultimate rules of action.
All things are interlocked, interlaced,
Interinvolved, interdenominated.
The pilot who pinched himself in his bomber,
Because only a year ago he was a boy in school,
Hardly realizing the magnitude of his change,
Was only one of a squadron,
A flying integer in a welter of heterogeneity.
Christ belonged to the Jewish race
But Chinese and Japanese speak to Him now.

The poet is a man of sense
Who handles the brightness of the air,
The viewless tittles he dandles,
Timelessness is his everywhere.

His blood is in the rose he contemplates
The blood of the rose reddens in his mind,
The poet is master of presences,
He is the insight of the blind.

Poetry is so mad and so kind
It is so majestic an inventive surprise,
Is it any wonder that in it
The spirit of man arise?

ON SHOOTING PARTICLES BEYOND
THE WORLD

"*White Sands, N.M. Dec. 18* (UP). '*We first throw a little something into the skies,' Zwicky said. 'Then a little more, then a shipload of instruments—then ourselves*'."

On this day man's disgust is known
Incipient before but now full blown
With minor wars of major consequence,
Duly building empirical delusions.

Now this little creature in a rage
Like new-born infant screaming compleat angler
Objects to the whole globe itself
And with a vicious lunge he throws

Metal particles beyond the orbit of mankind.
Beethoven shaking his fist at death,
A giant dignity in human terms,
Is nothing to this imbecile metal fury.

The world is too much for him. The green
Of earth is not enough, love's deities,
Peaceful intercourse, happiness of nations,
The wild animal dazzled on the desert.

If the maniac would only realize
The comforts of his padded cell
He would have penetrated the
Impenetrability of the spiritual.

It is not intelligent to go too far.
How he frets that he can't go too!
But his particles would maim a star,
His free-floating bombards rock the moon.

Good Boy! We pat the baby to eructate,
We pat him then for eructation.
Good Boy Man! Your innards are put out,
From now all space will be your vomitorium.

The atom bomb accepted this world,
Its hatred of man blew death in his face.
But not content, he'll send slugs beyond,
His particles of intellect will spit on the sun.

Not God he'll catch, in the mystery of space.
He flaunts his own out-cast state
As he throws his imperfections outward bound,
And his shout that gives a hissing sound.

A LEGEND OF VIABLE WOMEN

I

Maia was one, all gold, fire, and sapphire,
Bedazzling of intelligence that rinsed the senses,
She was of Roman vocables the disburser,
Six courtiers in Paris sat to her hats.

There was Anna, the cool Western evidencer
Who far afield sought surrender in Sicily,
Wept under the rose window of Palma de Mallorca,
For she thought fate had played a child in her hand.

There was Betty the vigorous; her Packard of Philadelphia
Spurred she; she was at home in Tanganyika,
Who delighted to kill the wild elephant,
Went Eastward on, to the black tigers of Indochine.

There was Margaret of Germany in America, and Jerusalem,
Of mild big eyes, who loved the blood of Englishmen,
Safely to voyage the Eros battlements of Europe,
Protectress to be of young and home, massive the mother.

There was Helen the blond Iowan, actress raddled,
Who dared learning a little, of coyness the teacher,
Laughing subtleties, manipulator of men, a Waldorf
Of elegant fluff, endangering to the serious.

There was Jeannette the cool and long, bright of tooth,
Lady of gay friendship, and of authentic song,
Beyond and indifferent to the male seduction
Who to art pledged all her nature's want and call.

There was the sultry Emma of West Virginia,
Calf-eyed, velvet of flesh, mature in youngness,
Gentle the eager learner of nature's dimensions,
Always to her controlling womanhood in thrall.

There was Sue, the quick, the artful, the dashing,
Who broke all the laws; a Villager in her own apartment,
She was baffled by the brains of Plato and Aristotle,
Whose mind contained most modern conceptions.

There was Maxine, a woman of fire and malice
Who knew of revenge and subterfuge the skills,
A dominator, a thin beauty, a woman of arts and letters;
She of many psychological infidelities.

There was savage Catherine, who leaped into the underground,
Her female anger thrown at abstract injustice.
And she could match her wits with international man,
A glory, a wreaker, alas, who now posthumous is.

There was Madge the sinister, who raged through husbands
    three.
She was somdel Groddeckian, a spendthrift of morality;
Existentialist that with men was dexterous
And would be in ten years after thirty, thirty-three.

There was a nun of modesty, who with service was heavy
And big with sweet acts all her sweet life long;
Enough wisdom she had for twenty ordinary women
Who percepted love as a breath, and as a song.

II

Where is Kimiko, the alabaster girl of Tokyo,
Living in bamboo among rustling scents and innuendoes,
To whom from Hatteras, the Horn, or Terra del Fuego
Returned as to a starry placement the sea voyager?

Where has time cyclic eventuated Vera
The proud noblewoman of Vienna? Among opera lights
She lived in a gaiety of possessive disasters,
Abandoned to the retaliatory shores of music.

Where is the naked brown girl of the nipa hut,
Under fronds, to Mount Mayon's perfect symmetry,
From the wash of the sea, looking from Legaspi?
Where in nature is this form, so brown, so fair, so free?

Where, who, sold into slavery in white Shanghai,
Walked and breathed in grace on Bubbling Well Road,
Subject to ancient sinuosities and patience,
Whose power was to represent unquestioning obedience?

Where is Hortense, the hermetically sealed?
Where is Hermione, haunted by heavens, who hesitated?
Where is Lucy, of bees and liberty the lover?
Where is Eustacia, of marionettes and Austrian dolls?

III

There were prideful women; women of blood and lust;
Patient women who rouged with scholarship's dust;
There were women who touched the soul of the piano;
Women as cat to mouse with their psychoanalyst.

There were women who did not understand themselves
Locking and unlocking misery's largess yearly;
Fabulous women who could not manumit the world
And babbled in syllables of the past and of money.

There were women committed to sins of treachery
The aborters of privilege and of nature's necessity;
There were the sinners in acedia of frigidity
Who negated even the grossness and grandeur of fear.

There were women without tenderness or pity
There were those more male than feminine men
Who rode the horses of their strident fury,
To whom subtle time made a passing bow.

There were independent women of society
Whose proud wisdom was their father's will.
There were mysterious women, Egyptian as a scarab
To whom scent and sound were a mysterious recall.

IV

Women are like the sea, and wash upon the world
In unalterable tides under the yellowing moon
Whose essential spirit is like nature's own,
To man the shadowy waters, the great room.

They come and go in tides of passion, and show
The melancholy at the heart of fullness,
Time crumples them, these vessels of the generations
Are crushed on the rocks as the green sea urchins.

They are the flesh in its rich, watery symbol,
A summer in July under the tenderest moon,
An island in the sea invincible to touch,
A refuge in man against refulgent ideation.

Women have gone where roll the sea bells
In the long, slow, the wide and the clear waters;
Their flesh which is our love and our loss
Has become the waste waters of the ocean swell.

They are the mothers of man's intelligence
To whom he is held by umbilical time,
And far though he roam, to treat with imagination,
He is brought home to her, as she brings a child.

THE VERBALIST OF SUMMER

I

The verbalist, with colours at his hand,
In the events and size of volant summer,
Thought at the sea's edge
How to wash the sovereign waters;

They were in a grandeur of the actual
And they leaped upon his eyes in tunes
That broke from island hills in blues
And flashed across the waves in mauves.

Is this the sea that balks my verbalism,
The mediator said, by flying hyalines,
Is this sea actual? Is this the real sea?
For he was the register of reality.

Or is it a chasm where old bones are rolled;
Forced peace with roaring rollers made
In the mad tangles of sea bells, fog, fate,
That specific for the forecast of our doom?

Or is it, while his nimble fingers flexed
The dawns, the sunsets of old centuries,

134

Awaiting the charms of elegance, of synthesis
Which ones to use, which tones to blend,

Is it the subtle messenger of nature
Hidden in the complexity of Psyche
That here appears, and are these bells, these shades
The temper of our mysterious complicity?

Was the water factual? Was it actual?
Did the eyes deceive, and did the senses drown
His too attentive blood in sudden frenzies
That made a music rare and rich in air?

He had the colour harmonies at hand
In the calm vegetation of his eyes;
In the aural cavities of carved sea rock
The rhythmic knowledge of neap certainties.

The verbalist, in all this wealth and scale,
Kept looking at the laughing, dappling water,
Hypnotic to the last, until his vision ceased,
And clearly then he saw that he was natural

To the menacing and the loving sea,
His intellect a super-imposition;
The verbalist sat in dells of verbalism,
As in a flower field, with music in the air,

And he made the sea level with his eyes,
No longer in the power of the imponderables.
All that mystery of the washing waters
Dissolved within his mortal reality.

II

The sea is mine, I am the sea,
For I am human if I am anything,

And I am the master of what I see,
Said to himself the cogent verbalist.

I am what I dream, the waters that I make
Are dreams of summer-scented splendours;
They are the dawns that stir the cormorants
Across the brightening ledges, and the gulls'

Elaborate conclamations. They are the loons'
Curious litanies, and the high osprey's cry.
I dream the waters, and I make the sea
Responsible; for I am what I take and make.

Subjective sea, sea of the deep sea-wish,
Waters of the blood that wash the world,
I claim the element of size and tone
My own, I make the sea, I am the sea;

Stylization of objects, fixation of dreams,
Compendium of imponderables,
Visionary screen on which is played
The magic games of the soul's reality.

So said the viewing verbalist
By the side of the ocean segment,
Mindful and instructed that he
Was, actually, foreign to this element.

And knowing that, where eyes were wide
To so many occult divergencies
His tones and colours were his own
The sundial sea was the plastic timepiece

As it had been when he began his reverie
And would be ten thousand years from now,
A living nature beyond the realm of art,
Try as poetry does to have it poetry.

III

I must begin again, the verbalist said.
The sea is always getting out of hand;
It is an addict of the summer's chances.
Those air-dances on the water, there, he said.

I must arrange the props, theorize
The distances and densities, defend
The subtlest of shifting elucidations.
The sea is a subject of some malfeasance.

He began to brim with violent intensities.
He saw a savage power, the ships broken up;
He had not thought so changeable a source
Could violate so many presumed images.

Accept the historians too; and of the weather
The weather-makers. And of sea-power
Its expositors and rampant individualists;
And place just so the salmon canneries.

The dullness of the ritual lighthouse keepers;
The thickening of the salt by centuries.
Typhoons, unseen, in the pacific sea;
Scott tying up to the Great Ice Barrier.

He began to think it was too much,
The subject proliferous and cantankerous,
No medicine to the subsuming spirit
Of order, gall to the onlooker, a tempest.

To be didactic about the ocean
Is to be a child, the verbalist said.
What does the child see in the ocean?
He sees the bright pebbles by the shore.

CONCORD CATS

The soft cat and the scratchy cat
Have milk in cold blue plates.
Then, in evenings, star-cool evenings
Equal to their reticence,
Emblems of independence,

These China cats, of black and white,
Will go on planetary pads
Uphill, where crouch
On eighteenth-, seventeenth-century
Houses, the graves of Concord.

By pious inscriptions
That antedate the Revolution
They see, through eyes cold and chaste,
The scratchy cat, the soft cat,
With humour old and Oriental,

That nature is meant for poise.
Battles, bloodshed, death,
Are men mirroring time,—
The stars blue, the night paling—
Are data. Imperviousness. Integrity.

ON THE FRAGILITY OF MIND

Mind is a most delicate evidence.
Not a soul has seen it yet.
And yet I think it is dense,
Although of great expense.

I suspect it of all trickery,
The master of the greatest paradoxes.
It is the historian of the world,
Crafty and foxy.

Old entablatures in Venice remind me
Of the mind of Tintoretto or Veronese.
These came to a watery nothingness
But for the golden paint I see.

I think there is no mind at all,
Perhaps, but only desires and faiths,
And the great capability of art
Which shows us forms, divine.

GREAT PRAISES

Great praises of the summer come
With the flushed hot air
Burdening the branches.

Great praises are in the air!
For such a heat as this
We have sweated out our lives toward death.

I used to hate the summer ardour
In all my intellectual pride,
But now I love the very order

That brushed me fast aside,
And rides upon the air of the world
With insolent, supernal splendour.

THE DRY ROT

The fine powder of the dry rot,
A new texture in the heart of the wood,
Crooked my finger as I scooped it out,
The rare old stuff of mankind's dream.

What to do with it? Pure contemplation?
Savage remorse? Illimitable irony?

A cosmetic for the face of Cleopatra?
Time in the blood is this pale substance too.

I mixed it with my sweat, and painted on
My legs and biceps bands of gold.
Sallying forth in the guise of Anthony
I sought the dusky land of sandy Egypt.

With a savage queen I lost remorse;
I threw my manhood down along a gold divan.
With a gesture of defiance and of loss
I met the meaning of the dry rot in that land.

Dig me up, then, in some later century,
Threading the pale sand through your thin fingers.
As you unclothe my bones in scholarship's delight,
Think of dialetic, imagination, the Rose of Venus.

THE SKIER AND THE MOUNTAIN

The gods are too airy: feathery as the snow
When its consistency is just the imagination's,
I recognize, but also in an airy, gauzy way
That it will capture me, I will never capture it.
The imagination is too elusive, too like me.
The gods are the airiness of my spirit.
I have dreamed upon them tiptop dreams,
Yet they elude me, like the next step on the ski.
I pole along, push upward, I see the summit,
Yet the snow on which I glide is treachery.
The gods are too airy. It is their elusive nature
I in my intellectual pride have wished to know.
I have thought I knew what I was doing,
Gliding over the cold, resisting element,
Toward some summit all my strength could take.
The gods are the fascination of the place, they escape
The genius of the place they make. They evade

The blood of our question. Imagination is a soaring,
It never allows the firm, inevitable step.
The gods tantalize me, and the gods' imaginations.
I am thus the captured actor, the taken one,
The used, I am used up by the will of the gods,
I am their imagination, lost to self and to will.
In this impossibility is my humility.

I saw an old country god of the mountain,
Far up, leaning out of the summit mist,
Born beyond time, and wise beyond our wisdom.
He was beside an old, gnarled trunk of a tree
Blasted by the winds. Stones outcropped the snow,
There where the summit was bare, or would be bare.
I thought him a dream-like creature, a god beyond evil,
And thought to speak of the portent of my time,
To broach some ultimate question. No bird
Flew in this flying mist. As I raised my voice
To shape the matters of the intellect
And integrate the spirit, the old, wise god,
Natural to the place, positive and free,
Vanished as he had been supernatural dream.
I was astonished by his absence, deprived
Of the astonishment of his presence, standing
In a reverie of the deepest mist, cloud and snow,
Solitary on the mountain slope: the vision gone,
Even as the vision came. This was then the gods' meaning,
That they leave us in our true humanity,
Elusive, shadowy gods of our detachment,
Who lead us to the summits, and keep their secrets.

THE HUMAN BEING IS A
LONELY CREATURE

It is borne in upon me that pain
Is essential. The bones refuse to act.
Recalcitrancy is life's fine flower.
The human being is a lonely creature.

Fear is of the essence. You do not fear?
I say you lie. Fear is the truth of time.
If it is not now, it will come hereafter.
Death is waiting for the human creature.

Praise to harmony and love.
They are best, all else is false.
Yet even in love and harmony
The human being is a lonely creature.

The old sloughed off, the new new-born,
What fate and what high hazards join
As life tries out the soul's enterprise.
Time is waiting for the human creature.

Life is daring all our human stature.
Death looks, and waits for each bright eye.
Love and harmony are our best nurture.
The human being is a lonely creature.

THE BOOK OF NATURE

As I was reading the book of nature
In the fall of the year
And picking the full blueberries
Each as round as a tear;

As I was being in my boyhood
Scanning the book of the rocks,
Intercepting the wrath to come
Where the hay was in the shocks;

As I was eye-drinking the waters
As they came up Seal Cove
With the eyes of my dazzled daughter,
An absolutist of a sudden grove;

As I was on that sea again
With islands stretching off the sail,
The real sea of mysterious time,
Islands of summer storm and hail;

As I was living with the love of death,
A concentrated wonder of the birches,
Passionate in the shudder of the air
And running on the splendour of the waters;

As I was a person in the sea birds,
And I was a spirit of the ferns,
And I was a dream of the monadnocks,
An intelligence of the flocks and herds;

As I was a memory of memory,
Keeper of the holy seals,
The unified semblance of disparates
And wielder of the real;

As I was happy as the ospreys,
As I was full of broom and bright afflatus,
As I was a vehicle of silence
Being the sound of a sudden hiatus;

As I was the purified exemplar
And sufferer of the whole adventure,

And as I was desire in despair,
A bird's eye in doom's nature;

As I stood in the whole, immaculate air,
Holding all things together,
I was blessed in the knowledge of nature.
God is man's weather.

Then I saw God on my fingertip
And I was glad for all who ever lived,
Serene and exalted in mood,
Whatever the mind contrived.

Then God provided an answer
Out of the overwhelming skies and years
And wrath and judgment then and there
Shook out the human tears.

COUSIN FLORENCE

There it is, a block of leaping marble
Given to me by an ancestor.
The hands that passed it held down ninety years.
She got it in the love-time of Swinburne.

This woman with her stalwart mien,
More like a Roman than a Greek,
Fumbled among old bags of rubble
For something indomitable that she could seek.

She saw the light of ancient days around her,
Calling in the hip-cracked hospital.
She chose at last. Then the clear light
Of reason stood up strong and tall.

With a pure, commanding grace
She handed me a piece of the Parthenon,

Saying, this I broke with my own hands,
And gave me the imagination of the Greeks.

I thought the spirit of this woman
The tallest that I had ever seen,
Stronger than the marble that I have,
Who was herself imagination's dream

By the moment of such sacrament,
A pure force transmitting love,
Endurance, steadfastness, her calm,
Her Roman heart, to mine, of dream.

I would rather keep her noble acts,
The blood of her powerful character, a mind
As good as any of her time, than search
My upward years for such a stone that leaps.

SESTINA

I die, no matter what I do I die.
Is this the sum of what man has to do?
There is no use to fly to be at ease.
Man flies, but knows not what he does.
It is in war you want to be in peace.
In Heaven, in Heaven I want to be in Hell.

The mortal span to find out Heaven and Hell!
No matter what I have to do I die,
The gods comply to cancel you to peace.
Before this then what is it man should do?
And after, does it matter what he does?
Will Christ-like Christ then put him at his ease?

Will will will him his own, a fabled ease?
Will, some say, is the whole road to Hell.
But man is bound to Hell whatever he does.

No matter what he does he has to die.
It is the dying that you have to do
Defies the hyaline lustre of the peace.

Despair has not the end in view of peace
Nor has desire the purposes of ease,
But action, while you live, is what's to do.
Thought is three crossed roads that lead to Hell,
Your thought is fatal and will make you die,
For thinking kills as much as action does.

It is not what he thinks, nor what he does
Nor what cold mystery of the Prince of Peace
Avails—no matter what I do I die,
May nothing, nothing put me at my ease
Except the reality of Heaven and Hell.
No one told me what I ought to do.

The scriptures told you what you ought to do.
They are unreasonable truth, and what man does
Believe when most he believes in Heaven and Hell.
That passes understanding, that is peace.
But sky-fallen man will not be put at ease.
I die, no matter what I do I die.

No matter what I do I have no peace.
No matter what man does he has no ease.
Heaven and Hell are changeless when I die.

'MY GOLDEN AND MY FIERCE ASSAYS'

My golden and my fierce assays,
My bold and sudden thunderbolts,
My costly lights, have led me all astray
And cut the ground from hope.

My will was avid, and my force
Commingling love and aspiration;
And my animal nature held
To auras of and beams of creation.

Wisdom comes in the short hours of night.
He came with the armed song of man
And I was destroyed in my light
In the coldest, daftest night.

For he said, the will is evil,
You live within a blind delight.
Violence is a fine delight,
Faith sang above the fight!

UR BURIAL

Reach me a blue pencil of the moon,
The double-reined rings from tombs of Ur,
The lyre, the javelins from Sumer,
'The Ram caught in a thicket.'

A gold dagger, a golden toilet case,
The gold helmet of Meskalam-dug;
Rein-ring and mascot from
Queen Shub-ad's chariot.

I will drink a narcotic rich and dark,
I will lie down by my master in his sarcophagus,
All our company will join in sleep
To serve the sun in a life beyond sleep.

SEEING IS DECEIVING

Having fed impalpably on the ineluctable
In yearlong passion of the unfathomed,
To realign the mystery of the ineffable,

147

Haltingly, haltingly, in lofty clashes,
Longsight glare, in total psychic refusal,
He would not allow his eyes to see the rose.

The rose that is with blood transfused,
The rose absolute, rose of all the world,
Flower of the flesh, flesh of the very flower,

Intruding organ of most rage-potential sense,
Rich rugged rose, the years in their dark make,
Sagest blossoms that can break and bless,

He would not in a high perfection sit,
Staring the day away in anarchy
That saw a heaven in imperfection's makeshifts,

And told the world in the etcher's fine escape,
The gardens clamant in the blare of June,
Birds precise, clouds fixed, grasses interlaced

With grasses as they are, and nature had,
But crying against perfection of the sight
He broke perfection with the inner eye,

Sombre, strange, amorphous, vast, replete,
And all his organs in the dark of time
Created the secret of an ultimatum.

ANALOGUE OF UNITY IN MULTEITY

A man of massive meditation
Is like a man looking at death,
Looking at death as at a bull's-eye.
He watches before he crosses the tracks.

Every day a man is in a box,
Hourly he watches the trains go by,

Opening and closing the wooden gates,
As one who is interested in the world.

The man who is massive in thought,
As it were of mountainous fortitude,
Whom decades have seasoned in male beauty,
Whose clarity is an age of harmony,

A man of intellectual power
Will not be killed by a ribboned artifice.
He is too full of deaths to be undone,
Death is his hourly communion.

　Who will say what the gate-keeper thinks?
　A man necessary to the metropolis,
　Comfort he knows, he keeps a comfort station,
　As decades pass on shining steel.

The meditative man, a power of eye,
Big shouldered, with the torso of a Jove,
Is master of a world of action,
An actor in a world of masks.

In complex thought he walks along
The most fearless man to be seen,
For with animal nature he is one,
Who looks ever death in the bull's-eye.

Far in the sky another eye
Beholds these creatures in their ways,
Indifferent to their differences,
A point of agate reference.

SEA-HAWK

The six-foot nest of the sea-hawk,
Almost inaccessible,
Surveys from the headland the lonely, the violent waters.

I have driven him off,
Somewhat foolhardily,
And look into the fierce eye of the offspring.

It is an eye of fire,
An eye of icy crystal,
A threat of ancient purity,

Power of an immense reserve,
An agate-well of purpose,
Life before man, and maybe after.

How many centuries of sight
In this piercing, inhuman perfection
Stretch the gaze off the rocky promontory,

To make the mind exult
At the eye of a sea-hawk,
A blaze of grandeur, permanence of the impersonal.

SAINTE ANNE DE BEAUPRÉ

The sun saw on that widening shore
Three hundred mothers with their daughters.
All dressed in white they followed slowly
Their mothers to the great stone doors.

O Bonne Sainte Anne
O Bonne Sainte Anne
The priest intoned upon the electric air.

The wind was bright from hundred years
And bright from off that distant shore
As slowly walked the maidens with
Their mothers to the cold church door.

A spectacle of heavenly imagery
And sun and wind upon the place,
The slow procession seemed a dream of time
While loudly through the air the cry implored

O Bonne Sainte Anne
O Bonne Sainte Anne,
Protectress of these children in their time.

Bronze on the hill, beyond the heavy Church
Stood dark the Stations of the Cross
In groves deep hidden from the sight
And from the brightness of such angelance.

O Bonne Sainte Anne,
O Bonne Sainte Anne,
Loudspoken tone upon white-flowing air.

Three hundred maidens beside their mothers
Slowly mount the tributary steps,
Entering to receive the blessing
Thought beyond all human ill.

O Bonne Sainte Anne!
O Bonne Sainte Anne!
An ancient spell lifts high in air!

And slowly came at last with blessings
And slowly down the prospect walked
And formed a slow and long procession
Of sun and wind and timeless innocence.

O Bonne Sainte Anne,
O Bonne Sainte Anne,

Protect the young in their extremity,
The wish of men who know all evil.

Such ecstasy had filled the hearts of many
And made so glad the light of tender eyes
It is a hurt that dream is not eternity,
And Bonne Sainte Anne not hunted down by time.

MEDITERRANEAN SONG

I like to see, when I am low,
The form I had when I was strong.

I charged the battlements of Carcassonne,
I climbed the white slopes of the tempting Etna

Startling the nymphs from the blue grottoes, the cool.
Transmitting new sanctions oblique and jocund

I threw my weight about the Mediterranean
In the true theatre of the antique world.

Almost a satyr in the woods, I was;
Almost Socrates in the market place.

TO EVAN

I wanted to give him some gift,
The breath of my breath, the look of my eyes,
I wanted to give him some gift,
Lying there so piteously.

I wanted to give him some gift,
Small child dying slowly,
With brave blue intelligent eyes,
His form withered piteously.

152

Only in the intelligence of those eyes
Where life had retreated for a piercing look
Was the enormous mystery justified,
As he inhaled the betraying oxygen.

I wanted to give him some gift,
A look from my look not to frighten him,
A breath from my haleness, my even vigour,
The same breath as his lonely breath.

Tenacious life in this little form
That will soon vanish from it entirely,
Unforgettable features of this little boy,
Do you mock my passion in your long passing?

I wanted to give him some gift,
Breath of my breath, the look of my eyes,
This is all upon earth, under heaven,
I can give him, a child dying, and I unwise.

Though I would menace the tall skies
And cry out as man has from the beginning
At the unequal fate held over us from our birth
I could not for a moment suspend this child's dying.

O though I would look into his intelligent eyes
With the world's weight of experience and despair
I could not mate the black look before death,
Nor seize the secret from the secrecy.

I wanted to give him some gift,
Breath of my breath, the look of my eyes.
Farewell, fair spirit. Fare forward, voyager.
I pass away silently and see him no more.

THE DAY-BED

I

It is green, it is made of willow.
I am baffled: I cannot think about it.
An obsession of twenty-seven years.
I am brutalized to look upon it.

The very form of love. Of time
The essence, which is memory.
The flash of light, and a long sleep.
This is the bed of day, and night.

No, but soft, but untold love
Arises. The very heart of love!
So long ago that suffering form
Slowly grew to death through pain,

Here on this very furniture.
It seems impossible. Time lies.
I do not see her lying there,
Great eyes, great gray-black hair.

I do not see that agonizing stare
That's deep through all my nights and days,
Substratum of the flying years;
The great pain without a cure.

II

Reality is a passing thing.
The Day-Bed lives, remains, reminds
Of the eternity of change
To this same, writing finger.

The emblem remains, bounteous gift,
The strange, pure gift of memory,

A blooded drench, a flushed presentiment;
And throngs and throngs of images.

Day-Bed of Life-in-Death,
That while my eyes shall change and see,
I look upon this furniture,
The not estranging imagery.

And summon up the love, and see
The very form and flesh of love
As it is with all mankind,
The loves long lost, the loves most near.

Who cursed the blood within the veins
Appareling day with source of night
Shall dream upon a lovely dream
Though the deep heart choke, and fight.

III

It is green, it is made of willow.
Lithe winds of Spring wave over it.
It is a new time and a new day,
New flesh here springs in harmony,

Laughs and tumbles and is gay.
Is gay! Is lithe as wings of Spring
And bends to nature as a willow
Triumphing in its green, cool stay.

Two lovers here electing unity
Flaunt eclectic idols in the day,
Consuming the great world of sense,
And laughing in its careless sway.

They sway. They laugh. And leaping
Loosen the mind from iron prisons,
Celebrating speeds of instancy
In vernal cells of intimacy.

Green and willowy marriage time!
Time of the beliefless flesh!
Time of the charges of the ruddy blood,
Joy that is swift and free, pure joy.

IV

Other years and other foils
Requite the ancient mysteries,
Persuading of some subtle balance
Between the losing and the winning battles,

Here on this very furniture,
Day-Bed of Life-in-Death!
A child plays in boisterous industry,
Truth off the old bones of mating.

Embroiled in fate he does not know,
Smiling mischievous and saintly,
Evidently impossible to quell,
The very future in his active eye,

The willowy Day-Bed of past time
That taught death in the substratum
Couches now the bliss of man,
A bright shape, a green new dream.

FORMATIVE MASTERSHIP

Never be happy until the golden hour
Creates its living death upon the instant
Just beyond perception: be happy then,
That you shall never understand that radiant
Mastery. It is gone as soon as known,
Known only as a breath of incalculable spirit,
Wordless peak except verbs prick at it;
Perhaps never known, but in this mating time.

Never be broken by the things of evil
That pass upon the years to true forgetfulness,
Giving themselves back to nature; man's evil
Forget, for in time forget you must,
Whether man wrought them from his crooked heart,
Or life imposed them in a cruel majesty
Impersonal and blind. Do not shake the fist
Or cry the brutal rage: time heals the time.

Baffled by instances of malice, keep
The calm of solitary imaginings,
That harmony inhere although the flesh be maimed,
Keep struggle pure with a white intent,
Revising possibility. Pare the naked nature
And in dark hours accept what fate is.
What toys we are to crippling chance
When victim, not the callers, of a savage dance.

Look upon the passing scene with tenderness.
All suffers change. The blight is in the air,
Within the lungs, within the light, within
The eye. Great nature is our master.
All our will and our flushed, enticed brains
Cannot unmake the world. Talk to the night
When the woods are deep, the stars alight.
Talk out the long instancy of mankind.

THE HAND AND THE SHADOW

The hand penetrated an abyss of shadow
When I came to the fullness of my being.
The shadow followed the hand in its search
Then leaped out of bounds, as if it were freeing

Itself from a compulsion and a tumulus;
And as if it wished to form new shapes,

And command a function of new elegance,
And as if it had the dignity of grapes;

This shadow that would escape the lordly hand,
And would break the bonds of its condition,
And be a fruit just at the point of bursting,
When meaning leaks from commission in emission;

This shadow now transfigured and as of flesh
That has in it the blood of memory,
And as of all blood that has in it action,
And rises in imagination's cheek as primary,

This shadow is now the grape at an exquisite point,
On a day of Autumn, the fullness of the year,
The hand has become something departed from,
And sits upon a symbol, as a tear

Breeds upon the eyelid of a natural action,
And bursts upon the fields of contemplation,
And, rich with death, and then with resurrection,
Sows a radiance round about the nation,

Then, as a glory pens the eye of man,
The shadow-semblance of a heavenly sun,
Comes back to humble energies and potence,
And the shadow and the hand are one.

WORDS

First a word was fuzzy, and was nothing.
It had all India in fee.
It had to do with illusion's very self
And nothing to do with me.

Then a word was mighty, and flew away,
An eagle on a rocky shelf;

It glowered over our America,
Depleting the inner self.

Finally a word was elegant and sheer,
Like the fabric of a stuff
Durable with the fate of things,
When I had lived enough.

Then I had a total myth
Sounding on an eternal shore.
It was the world-memory alone
When I was dead and gone.

ON A SQUIRREL CROSSING THE ROAD
IN AUTUMN, IN NEW ENGLAND

It is what he does not know,
Crossing the road under the elm trees,
About the mechanism of my car,
About the Commonwealth of Massachusetts,
About Mozart, India, Arcturus,

That wins my praise. I engage
At once in whirling squirrel-praise.

He obeys the orders of nature
Without knowing them.
It is what he does not know
That makes him beautiful.
Such a knot of little purposeful nature!

I who can see him as he cannot see himself
Repose in the ignorance that is his blessing.

It is what man does not know of God
Composes the visible poem of the world.

. . . Just missed him!

CENTENNIAL FOR WHITMAN

(Amimetobion, not Synapothanumenon)

I

What shall I say to Walt Whitman tonight?
Reading him here in the springtime of bursting green,
Foreign from him, held by the same air he breathed of the
world,
Looking at night to the same stars, white and radiant,
Obsessed with a kindred obsession, at a dark depth,
Inheritor of his America maybe at its great height,

I praise him not in a loose form, not in outpouring,
Not in a positive acclamation of frenetic belief,
Not in the simplicity of a brotherhood, such peace,
And not in the dawn of an original compulsion,
But speak to him in the universe of birth and death.

By a Spring meadow I lay down by a river
And felt the wind play on my cheek. By the sunlight
On the water I felt the strangeness of the world.
Prone in the meadow by the side of the fast brook
I saw the trout shooting his shadow under the willow.

I sank into the mystical nature of memory
And became my beginning. I was one with strong nature,
At the heart of the world, with no need to penetrate her.
In the sheerness and the elegance of this feeling
I destroyed time and dwelled in eternal pleasure.

The vastness of the aim of human nature
Yielded to ease and immediacy of comprehension,
Such is the rarity of the mastery of existence
In the ethereal realm of pure intuition,
Within the subtlety of perfected spiritual balance.

II

What shall I say to Walt Whitman tonight?
Nothing that is not myself. Nothing for himself,
Who spoke the golden chords of a rough soul
Deep below the meeting of the mind
With reality; his words were a mask of the true soul.

I grew up among animal pleasures, hot in sense,
And fought off the lofty reaches of the intellect
As one knowing the soft touches of the night,
Running on the Spring freshets in delight,
Joyful and serene, not to be overcome or quelled.

Then dramatic evil like a blight overcame me,
The dream-like character of eternal knowledge
Was brought in earthly bondage; knowledge of death,
Our old enemy, appeared with his powerful will
And laid waste the garden of my green seeming.

The years began to whirl in a worldly ecstasy
Fulfilling some dark purpose confronting the heart
Of things, and I was loosened to flesh and mind,
Torn asunder from essential unity
And would wander the world in fateful duality.

This was the knowledge of good and evil,
This was the certainty of actual death,
The powerful hold of an ancient, fallen state,
The battering ram of time on the bones and eyes,
The new reality of the unredeemed mankind.

III

What shall I say to Walt Whitman tonight?
I look not upon the world of facts and figures
But in the heart of man. Ineradicable evil
Sits enthroned there, jealously guarding the place,
Only held at arm's length by a comic attitude.

Laughter at the sun and the moon, at the tides,
Laughter at the comedy of the etrnal struggle,
And at the institutions and society of mankind
Laughter, I celebrate this tonic attitude,
And go as far as that for the sake of intellect.

And run on bitterness and corrosive pessimism
Standing under the glaring eye of antique satire
And range the fields of powerful condemnation
As one who allows himself such pleasures,
A beast engaged, knowing the gates of escape.

New bombs, new wars, new hatreds, new insecurities!
Man has become the victim of delusions
Thrashing his brains in energies of misaction,
Lost in tribal sin, ready to destroy himself,
Defenceless against all natures of monstrosity.

What shall I say to Walt Whitman tonight?
Give us a share of your love, your simplicity,
The large scope, the strong health of the soul,
Love be our guide, and love be our redemption,
Love make miracle, animate us now.

IV

Love come upon us when the willow bends,
Love come upon us at the child's upturned face,
Love recapture us in the market-place,
In churches, slums, on mountains, in the fog,
Love be with us in the hour of death.

Love be with us in the pang of birth,
And throw out hatred, envy, pride, despair,
Be joyful at the time of the tall daffodil,
Be rampant as the legendary lion,
Be meek and sweet, and sure, so love be here.

Love that is swift creator and saviour
Bless all the infants and the old men,
Bless the middle kingdom of the workers,
Love come in the soft night, in the sensual day,
Let our airs be soft flower-lofts of love.

What would you say to me, Walt Whitman, today?
Is there anything you can give me but your love,
That total devotion to comprehension of the word?
It is not the forms you evoked, these are changed,
But the force you spoke with, the heart's holy rapture,

Your knowledge of the changeless in birth and death,
The merit of man in his eternal suffering,
Your love of the stars, of valour, and of doom
That I would say to you, Walt Whitman, tonight,
That you could say to me, Walt Whitman, today.

SOUL

I

The flow was toward the land,
It was a music of the altar deeps,
Impersonality of the proving waters.
Soul was coming to the landfall steeps.

The Soul was coming in it seemed at neap.
It made a faint halation in the air
As if a white, mist-fitted presence brooked
The high rocks with some supernatural stare,

Mysterious essence cleaving to the bushes,
That gave a strange elation, the good of man,
Subtle benediction like a waft of spume,
A sense, and then it seemed it was not there.

And came again before the mind had gone
Back seaward where the spirit was, that source,
And roamed a dream-like presence over earth
As if it had been always there, for man.

II

The Soul was fugitive, but it was fronds
Of palms from some soft South, palm-tones
Weaving the tides with rhythms warm and rich
That crept within the blood-dream like old time.

Then it was a presence of the memory
Walking on the land in heavy imagery
And gave so supernatural a sign
The heart rejoiced and had no more to fear.

The flow was massive as toward a midnight,
Sea-music deep as dreams of history,
And then it went away upon the air
Like silent boats pushed off the full-tide shore.

A holy feeling came upon the beach
And penetrated inward toward the woods,
Whether the Soul was far slipped out to sea,
Or whether a presence held the mind again.

III

Then it came upon the dappled waters
Across the farther islands, inward coming,
And seemed to be, in truth, the actual Self,
Safely stationed in the standing man

Who, breathing on the mystery of spirit,
Partook of insight, becoming a living Soul,
And was one with that going, and that coming,
The Self bound up within the eternal Soul.

Father of mankind, Soul that comes from deeps
And flows upon the land in fullest tides,
It is in mortal sonship that I stand,
With the heart's mystery in either hand,

And wait upon the judgment of the hours
That nothing should make clear this ancient praise
But ever keep, in mind alive and deep,
The dark intuitive presence of the Soul.

FABLES OF THE MOON

To keep an eye on fables of the moon
While feet are walking down the street
Is not the least a man can do,
And may be most.

I think of an old secret.
I walk upon the mart, and swear
It is a man's part so to do,
But not with my whole heart.

It is needed to be quick
And deal the world out with a will.
Gratuitous to think
On a sudden, summer whippoorwill.

SALEM

Madly she came, and stood upon the lintel,
With a look that you should never see.
And madly she handled the frail air,
Unimpeachable, a seer.

And with a visionary grace inwrought,
As who was moved by powers,
She was the representative
Of a teeming demonology.

On the hill there is a gallows
Still in the form of a blasted tree.
If you will look through the thinnest air
You will see what you will see.

THE RETURN

Still marvelling at the light,
Impersonal, on the mountain peaks, a halcyon
Glow; it strains to me,
To the last intimacy.

Then, quick to seize on intuition,
I thought I knew; now I know
I do not know. Time has refracted
Ineluctable meanings.

Now, the sight is more satisfactory.
Decades make us mountainous.
Life did not know what time could do.
My long light streams out to you.

THE GIANTESS

Idleness gilds an encompassing giantess
Who lies upon the heavy land of torpor.
Her yellow flesh, empty in indolence,
Poisons desire; she is mordant and depraved.

I have walked upon the golden stones of time
And fallen on her lurking on the hills.
My senses shrink within her slow, brutal languor.
She holds a sullen summer smile of lassitude.

She draws the spirit down upon the stones
And makes it as a gorged and sleeping snake,
Monster of disuse, elongate in her charms,
Where time is working its worn dreamlessness.

She moves in slowness of a long caress
And mauls the future. Languid and primeval,
She sucks evil away in one vast innocence,
And holds her victim mindless like a night.

Such hours of leaden torpor in her arms
Control an elemental misery;
While senseless of her flesh, her sensual power
Corrupts the longing and the love of death.

She is the mood of land-gone sailors when,
Their copra hoisted out at hot Copan,
They seek the thatch of love, and hope, but find
The girls are gone, never to return again.

THE WISDOM OF INSECURITY

The endless part of disintegration
Is that it will build again;

167

Of a robin,
That he will become a memory;

Of a hand-great August moth,
With eyes in his wings, so fine

As to represent himself as a fable,
That he will be tried again in a poem;

Of a petal-departed rose,
That it is its loving grandparents;

Of the evil of man to man,
That it is the tenacity of mankind

To put it out of mind in the long time
And commit to the heart new fidelities;

Of the broken music of silence,
That it will be brought together,

Renaming poetry; of the day's colours,
That they make the painter's paradigm;

Of immeasurable fallibility,
That it becomes art, by authority;

That the disintegration of time
Becomes the enrichment of timelessness,

So that nothing is destroyed, but finds
Its truth in the eternal mind;

That death provides nowise no escape.
Man becomes some other shape.

I must have come down from Adam
And am his metropolitan.

The leaves are falling and the rain.
Mystery is palpable. Sameness ever the same,

Ever different, where I look
Is through the fish to the fish-hook.

The strangeness of the poet's dream
Will set what is, not what seems.

When you see destruction itself
You see form out of the formless

As I recall in the Campo Santo, at Pisa,
The work of medieval Giovanni

Who painted an old man, prone and dead,
An infant, the soul, springing from his head.

SUNDAY IN OCTOBER

The farmer, in the pride of sea-won acres,
Showed me his honey mill, the honey-gate.
Late afternoon was hazy on the land,
The sun was a warm gauzy providence.

The honey mill, the honey-gate. And then,
Near by, the bees. They came in from the fields,
The sun behind them, from the fields and trees,
Like soft banners, waving from the sea.

He told me of their thousands, their ways,
Of pounds of honey in the homely apiaries.
The stores were almost full, in Autumn air,
Against the coming chill, and the long cold.

He was about ready to rob them now,
The combs. He'd leave them just enough to keep them.

I thought it a rather subtle point he made,
Wishing Providence would be as sure of us.

SUMMER LANDSCAPE

The pine with gray tips is moist still.
Peace pervades the scene. The standing pool
Has the green mantle, a spreading tone.
All the walks are full of time.

Vigour stirs the rose in thousand petals
As the fountains play; the frogs are placid.
It is a moist July. Rich the blood.
The snake will hold his poise, then glide away.

I have thought of pastoral simplicity
As flesh without a mind; it is luxury
Increasing summertime, as strokes of sleep
Accomplish better looks when birds are winging.

Swallows, for instance. There are swallows playing
About the barns like brightest minds; their eyes
Are quicker than my eyes; they make my mind
The quickness of their turning. I, their quick.

Cordwood stands beside the heavy vegetables.
Rank growths encroach in lower meadows; wild-
Run rough soil keeps a flaming need
Of growing trees, young pines forcing the landscape.

Everywhere is ease; everywhere is peace.
If an emotion comes, then put it back.
It is the stillest drama of stillness itself.
Poise is delicate, dedicating this instruction.

Now waves of the clover-ether triumph.
Space is a time of abundant silence.

The supple bees abound. The squirrels play.
The moonlight will be as good as the hot day.

ONLY IN THE DREAM

Only in the dream that is like sleep
When time has taken the measure of live things
By stark origination
Is mankind redeemed.

Only in the melancholy of the music
Of the midnight within the blood
Comes the fulfilment
After faring years.

Only in the balance of dark tenderness
When everything is seen in its purity
Do we penetrate
The myth of mankind.

Only in the mastery of love
Is anything known of the world,
Death put aside
With pure intent.

Only in the long wastes of loss
Comes the mystical touch on the brow
That triumph grow,
Insatiable, again.

NOTHING BUT CHANGE

We saw nothing but change in all the ways we went,
Nothing but time weaving wind, weaving the willows,
The tall buildings taken down and put up again
As we looked at them; the bridges made again.

We beheld the same of persons, nothing but change,
A kind of personal constancy we thought it to be.
Inured, we finally felt when somebody died
It was part of the flow of time, the nature of ourselves.

Then time itself began to get the upper hand
As our bodies, falling to the stress of air,
Left us to ridicule; Plato and Christ were bare.
A life-time of thought could not get through the flesh

And all the pledges of superlative intellect,
The rich, sensuous breasts of memory,
Fictions of the spirit, whips of the blood
Met denial, the unanswerable, God-given death;

The veritable and very rich death,
Without a poem, riches itself,
And when the gaudy play was enacted,
It was all a roundness, all a poetry.

THRUSH SONG AT DAWN

Bird song is flute song and a glory
Of the morning when the sun unascending
Holds his other glory of mentality

And the dawn has not the mental mockery
But the birds from sweet subconscious wells
Pierce through all barriers to sense,

They send and giving sing divinity,
So sweetly charged with subterranean meaning
They are like angels in the morning

Come from ancient time, a fast enchantment,
To bless our mortal songless weakness
And trail a vocal glory all the day.

I would not be a bird, but I would hear
Deep in some lost purity, beneath the mind,
There in the sweet, dark coil of time,

As in a mother, the thrush as saviour,
And a sovereign mediator; or any other
Lung-red singing: richness propounds confusion,

That pleasure that will never cease to be
Where we are played upon without a fault
By magic tones we love but do not have to know.

THE VOYAGE

To make a headway against the tide,
The tide-rip and the windy afternoon,
The skillful sailor, salt consigned,
Impressed the canvas to new subtlety.

The engine off, a sporting gesture.
Found him moving aft, heading fore.
Heading forward, we go backward
Where we were. The day is pure

And time seems stopped somewhere.
The sounding shore, we passed far back,
As ancient as Circean chant
And lure, comes to sound once more.

Now the consternated sailor winces,
His skill debased, the wind long gone.
The open ocean, a lolling torpid giant,
Occasions nothing but philosophy.

He turns the engine on again
As a last resort. Man cannot endure
To be going backward going forward.
The sirens wail in the rock's sonority.

OFF SPECTACLE ISLAND

Seals and porpoises present
A vivid bestiary
Delightful and odd against the mariner's chart.

The sea bells do not locate them,
Nor lights, nor the starred ledges;
We are unprotected from their lyricism.

They play in the blue bay, in day,
Or whoosh under the midnight moonlight;
We go from point to point where we are going.

I would rather see them playing,
I would rather hear them course
Than reach for Folly from Pride's Light.

THE SEASONS

Spring was never his fort. Suspect at first,
That is, at the first blush introspective, not first
When voluptuary to the inner arm, when durst
That smooth, round, and small quality, that nursed
Times earlier even to the earliest sensory charm,
As Burke would have it that way to be beautiful,
Not as the gowning roundness of the flesh growing,
Not absolute, that is it, and not a metaphor;
It was green, it was there, seen, hardly a
Thing seen, a deposition of some kind and style,
Overweening, but not a thing of criticism,
A lean-to in the gingery footstep forestation.

It was a trouble to the port. Whole scopes
Range in the vanguard of the watery ropes;
A voyage is many after-thoughts of hope.

Heroes have testy decks, not caulked in dope.
There is something lost in this misuse,
Minimal asphodels, worm-fat robins in the parks,
The structures of the erstwhile hyacinth,
And heavy gear of many a heady mallow,
No doubt. He cast his wits upon a sea
To find out. The nets trailed slime, green ooze,
And came out glistening in bright free gesture
Mounting, mounting time, genial Spring thrown out.

SPRING MAN

Summer was the problem of the universal,
The reductive mint-julep, seductive midnight recall;
The sweat upon the face; the glove homing the baseball;
Or the hand on the moonlit tiller, quiet as all;
No judicious panting after Heraclitus,
Nor fractions of Plato should burke the days' stale,
No frantic cry to Christ in the apt, higher stones,
Nor self nor soul lost or saved in mobbing thought.
Illumination. Without the need of intellection,
Imagination. Without the framework of doctrine,
Creation. Full as sense, time's melon, the scandal
Of elation. Love's game, matutinal, primeval.

The sun was the master of simplicities. The trees
Were a happy garden state, a golden grease.
He had the notion they could not decrease,
His flesh thus unguent and liquor, fat fleece.
The time of the year without fear, the pure time,
The time of the melon, and apple, the luscious pear
Still without a hint, burdening the fleshly air
Of anything but rich ease, heat weaving in air.
Not the mind, the flesh. All wholeness and wantonness
Was to human kind. The meadows with yellow sweetness
Would ever shine. Dance of the air-bees and blue brids,
The sea a brine of blood, ever without a mind.

THE MAN OF SUMMER

Autumn was the heavy mood of the human,
The heavy clutch of memory of everything known,
Lost and regained now in slow, gravid time:
He entered it as entering a world-mine.
He put his chances to this plunder. A long
Glare of introspection sealed the days in slow
Feeling, the half-closed eyes had seen too much,
And all was the sombre heaviness of betrayal.
Knowledge. The deep knowledge of the Autumn coming
To death's edge. The certainty of the decreasing future
Pledged imprisonment, that no backward look to tendrils,
Or vegetation, came without a harm or hurt.

It is the mind of the oldest Schopenhauer.
Recognition was within the hollows of the hour,
An inability to change, man without power
Empowered to see the truth; then not one flower
Would dance: not a thing would grow to fiction;
Not the salt sea dazzle in dalliance;
Not mountains inspire, nor metaphysic beckon.
The sufferings of man were every realization.
Autumn days! They hang upon the spirit like the flesh,
An exaggeration. Down to doom then man must go,
The old, the gay. See how the long centuries dwindle.
Do not stay. Knowledge is without power. O do not stay.

THE MAN OF AUTUMN

Winter is for Eskimos a castle of content,
Facetious in the furor of the snow; something blent,
Impure, high-pitched, finally an integument,
Destructive of the malice it engenders: embalmment.
It was what nature slyly left, but gets in full.
It is the mind's best, last chance and terminal,
The gaff of laughter, derision's joy, that icy joke
The deftest handle: satire of intelligence itself.

Zithers! Winter lacks the opulence to know.
It withers. It is the coldest, nicest predicament
Of dithery sheens. It is its own blight and refuse.
Whither? Winter is the purest intellectual dream.

It is significant on coldest winter nights
Love studies hottest: the animal has nothing quite
So straight an answer to that universal, white
And comic aspect of the natural blight.
And man will keep his courage, his provident esteem
Where most the elements stay mercury in spleen.
So in the deadness of the winter time
His blood, illogical, will seek the slime,
The melon, or even the withering of the vine.
Time play on. He cannot be dissuaded of his place
In the sun. Man is his own invention,
The Son of Man his timeless election.

THE NOBLE MAN

An Olympian before the Olympics, that inhuman backdrop,
The possible man, it is him I posit, him preempt,
The not incalculable master of the braced mountains.
He does not have to lose himself in them. His spirit
Creates their newest hazard as idyllic only.
He has confronted man, Olympian of the dark mark.

Rage is the substratum of pure animal poise;
A strenuosity, exquisite, roves in the controlled eye;
Multiple actions, muted, gaze upon the world;
Leaping tensions exhale the formidable and graceful;
Masterworking modulations enter every day;
Oval kingdoms quell the old volcanoes.

The Olympian looks at the Olympics profoundly,
In fervour. To him they are fractures of the Alpine.
Their height is modified in his height. Hurls
Fireballs out of himself when he wills it. Towers

177

Over his own music, lofty, sensitive, possessed.
The Olympian is master of his own mystery,

Mystified in his own mastery, recognizing
His relation to the real, the impersonal mountains.
He knows his own reality and that of the world,
The most human one, testator of furor and quiet,
He fears not, having feared all, loves altitude,
Gives off love's avalanche, mountainous man-make.

Fear that undoes many straightens him;
Contemplation is his burly action;
Round is the eye of his understanding;
Fear of death has not shaken him;
He looks upon all things with fervour of tenderness;
His laughter is ready. He has overcome pride.

An Olympian will not drown in the rain forests
Of self pity. Life gives him its own mountains.
Active love is the mark of a noble man,
By cynicism, by ascetiscism not betrayed. Nature
Works in him her balance. A bright perfection
Plays around the brows of the Olympian.

THE FORGOTTEN ROCK

Tawny in a pasture by the true sea
Ship-shaped it stood, the never realizable
Rock.

It was the awe placed on this natural object,
When looking at one thing, another world was seen,
He remembered.

Even this rock, sun-stroked, high-fashioned,
Peopling dream, enticing him completely,
Was never itself.

178

With such subtlety the blood, blue-watery source,
Leaped from one thing to something other.
A double escape,

A paragon of sight was posed upon
The blue rock; the excitation of the world
Glanced off.

This was nothing but the world itself;
A voyage in blue distance under tawny charges
Vividly known,

As a ship like a jewel of sense, come to
With boundless appetite, is yet unknown,
And is forgotten,

As we forget time that is thoroughly gone,
But was so certainly there. We annihilate time
To remember,

And what we remember is the duality of time.
The rock was never known as it was,
But as we are.

It had a cave with an obscure dome
A hawk flew into heading home,
Killed instantly.

ATTITUDES

IRISH CATHOLIC

After the long wake, when many were drunk,
Pat struggled out to the tracks, seething
Blinded, was struck by a train,
Died too. The funeral was for the mother and son.

The Catholic music soared to the high stones,
Hundreds swayed to the long, compulsive ritual.
As the mourners followed the caskets out
Wave followed wave of misery, of pure release.

NEW ENGLAND PROTESTANT

When Aunt Emily died, her husband would not look at her.
Uncle Peter, inarticulate in his cold intelligence,
Conceded few flowers, arranged the simplest service.
Only the intimate members of the family came.

Then the small procession went to the family grave.
No word was spoken but the parson's solemn few.
Silence, order, a prim dryness, not a tear.
We left the old man standing alone there.

AN OLD FASHIONED
AMERICAN BUSINESS MAN

I asked no quarter and I gave none.
I fought it out until eighty-one.
Intelligence guided my efforts,
Cynicism dictated my business reports.

I played one off against another,
Keeping my head well above water.
I transmuted my blood to steel,
Early conquering the realm of feeling.

I taught ruthlessness by example
And hard work and steadfastness.
Now that the hated grave approaches
I wish for the love I could not give.

A YOUNG GREEK, KILLED IN THE WARS

They dug a trench, and threw him in a grave
Shallow as youth; and poured the wine out
Soaking the tunic and the dry Attic air.
They covered him lightly, and left him there.

When music comes upon the airs of Spring,
Faith fevers the blood; counter to harmony,
The mind makes its rugged testaments.
Melancholy moves, preservative and predatory.

The light is a container of treachery,
The light is the preserver of the Parthenon.
The light is lost from that young eye.
Hearing music, I speak, lest he should die.

PROTAGONISTS

To the man with his jaw shot away, blood-badged,
As he falls out of the sky to earth or sea,
Existence is lethal, and then it is not. Then

It is as it was in the beginning, and ever shall be.
    His mother sent him a Cross on Easter Day.

Last seen on Widener steps, holding a *summa cum laude*,
Otis Peabody, head large and capable as the stars,
Understood and understands the law.
A flak burst in the head destroyed the whole of Harvard.
    O you I admire—luck is a fatal absolute.

Here is a gay one home from the wars, Patrick O'Shaughnessy
Keenness flashing from the Distinguished Flying Cross.
Of forty bombing missions over Europe, pilot.
He teases the room with a neat card trick, and
    Homer never had a hero half so grand.

But I see a man in blue denim walking, walking
Through the halls of conscientious objection,
Because he took Christ seriously, immured.
A literalist of the imagination! who
    Believed do unto others—Thou shall not kill!

A SOLDIER REJECTS HIS TIMES ADDRESSING HIS CONTEMPORARIES

Think no more of me
For I am gone
Where ages go,
No touch of snow
Nor sing-song bird sip song
Alters meadow.

You are whelmed in manyness
By days totally taken
By doing taken up

Summer and winter,
Plunger of fortune.

But I am hidden
And dropped out,
No more I count.
Flesh was fresh,
Time was evenness,
Death is stiffness.

Think only of the living,
It is the lips of wisdom;
Of wars perpetual;
Even of steadfastness
In man's misery.

'BLESSED ARE THE ANGELS IN HEAVEN'

Blessed are the angels in heaven
We are not with them on earth
And blessed is their rationality
Denied us from our birth

And if any can say
In our writhen, human condition
Proud, red words to make us gay
Let him say them and let him be

A man of courage in a broken time
And a great singer in the evening
The master of our irrationality
And lover of our writing

And the angels in heaven who are blessed
Will not hear him calling, calling
For they have nothing to do with us
And bind us to our barking heritage.

VILLANELLE

Christ is walking in your blood today,
His gentle tread you cannot hear nor see.
He tramples down your militant wish to slay.

The whelped deaths you dealt in your war's day
Arise howling, they will never make you free.
Christ is walking in your blood today.

You did it easily in the heat of the fray.
You did not know what you could do, could be.
He tramples down the massive wish to slay.

You are the front and fore of passion's play,
Of deepest knowledge you have lost the key.
Christ is walking in your blood today.

To kill is one, is not the essential way
Of action, which you then could not foresee.
He will wash the welling blood. Not slay.

A child becomes a man who learns to pray,
A child-like silence on a moveless sea.
Redeemed you may be of the will to slay.
Christ fermentative be all your blood today.

LIFE AS VISIONARY SPIRIT

Nothing like the freedom of vision,
  To look from a hill to the sea,
Meditating one's bile and bible: free
  From action, to be.

The best moment is when
  Stillness holds the air motionless,

So that time can bless
　　History, blood is a caress.

Neither in landwork nor in seawork
　　Believe. Belief must be pure.
Let the soul softly idle,
　　Beyond past, beyond future.

Let it be said, "A great effulgence
　　Grows upon the sandspit rose.
A rare salt harrows the air.
　　Your eyes show divine shows."

FORTUNE'S MIST

The mists of fortune
Blow with wayward cadence
Within imagination
Making chaos into sense.

Man dreams his tribulations
Strict and articulate,
As Lincoln by a fire
Wrote history on a slate.

If not by absolutes,
At least by artifice
We cast our devils out
To seek an angel's face.

Strength in the hand and mind
Holds to the central doom,
Working reality
In coping with a tomb.

But for imagination's power
Shaping the unseen

Fortune annihilates us
As if we had never been.

A heavy head of snow
Tumbling from a hedge
Is March's triumph; its revision
Is the poet's pledge.

Art gains in time
For it was deep in mind,
That rounded the waters,
And shaped the wind.

YONDER

I held Europe in my hand
Like a jewel that is implacable.
When I became most interior
I fled to my native land;

I returned to America, singing
Poets dead, long-living causes;
I loved them all. About them
I felt hypocoristical.

When the lovers walk on the Danube
And the thinkers vie in the park
I give my hand to the Big Horns,
I seek the Pacific in the dark.

AUTUMNAL

Action is a pain of being,
    Being is a rain of knowing,
        Knowing is a sense of self,
            Lovely the day

Thereof. Would you have it otherwise?
    A peak of ants of little size
        Obstructs me on the sidewalk,
            All clamoring talk.

The statues are lying around on the green
    Seeing what they have seen
        As lovers pass in the grasses,
            Elemental, massive.

Big clouds and little clouds cohere
    In arguments this time of year,
        Unclassified; they are lapses
            Floating, dappled.

Dogs are in and out. And bees
    Are falling into cider granneries
        Fat as unbuttoned fops.
            One drops, another drops.

We have been living the full year,
    It is still full, it is here
        In the late recline of sun,
            A red grand one.

What is going on beyond
    I have not found, am bound
        To the love of the unfound
            Beyond, but here.

THE SACRIFICE

Every poet is a sacrificial spirit,
Every song he sings is given
By special election in heaven
So that men may bear it.

Each toils, and throws his life away,
Gay as a boy tossing his cap up
For whether of tragic things and heavy
He lives in the senses' gaiety.

Like some drunken bee plundering flowers
Drunk with his gorgeous nature
He gives a golden summer afternoon
Its fiction, intellectual and sensual.

LUCUBRATION

Rain upon the window, rain upon the day.
The vastness of the stars ever upon us.
Fluid ripeness where the grasses totter,
A chill descending with another year.

Savageries of youthful indignation
Intimate the will: their lustful excellence
Flows out upon a red, eventual brightness,
All the future leads them up to dragonfalls.

O but the stealth and silence of the skies
And starlight far and starlight still and cold
Brings on the close communion with defeat,
The forceful draft of spirit's nightlong questioning.

And O again the children's vision, all delight,
That quick perfection seen as senseless innocence,
Against the rule of reason, regimens of solid day,
Perplexities of sensibility, the soul caught in a thicket.

Rain, the giving rain, leaves' quick departure,
The threading time, and blanketing, cold metaphor,
Time's dualism, where we heft the airy gait,
Where stares the negative stone, there Hynos waits,

Perhaps, some final, psychic resolution,
The tongue of fire upon the fuming water,
The eye of purpose in the lust of brightness,
Song's mastery upon the starlit tomb.

IN AFTER TIME

In after time, when all this dream
Becomes pure dream, and roughest years
Lie down among the tender grass,
And spring up sentient upon the meadow;

In that after time of great-born Aprils,
Beyond a century of tatters and of malice,
When love has thrown out fear and madness
The eyes will see the sun as wonder.

In after time, when rage and chaos
Lose their sovereign force, new dream
Will lift the shining life to spirit
And mate the make of man to merit.

Then shall holy summers come; then laughter
God-like shake upon a dewy morning;
Then fullness grow, big with purpose,
And man shall know again his richness.

A TESTAMENT

It is what I never quite understood
About my formidable day
Was the truth, the trust, and the good
Where the final values play.

It was the ungraspable part
As events and crises interfered

Was the bold quest of my art,
The blood, the crystal tear.

It was the lack of direct mating
With authority, politics or powers
Gave thrust to loving and hating,
Allowing the purest hours.

It is the world's inability
To nourish my senses whole
Brings on the subtle utility
Of the total strength of soul.

I sing of something so far and deep
Ages will find it clear,
Love that is a grasp, a leap,
And faith, landfall of fear.

Henceforth this testament
Of the struggles of my bone and time
Is made; for those meant
Who cast the spirit, seed the sublime.

REQUEST

For Dame Edith Sitwell

Will you write in a new book
Just your name,
In a new book, for me?

Will you write it plain,
In a new book, your name,
Just your name for me?

Will you make it an emblem
Of seventy years of joy,
Of pain, just plain, to see?

So that when I see this name
I see the turn of your hand,
The turn of the years I see

And think of the musical poems
That have made you swift and free,
The long waves of the sea

You caught in a life-time of
Rhythms to be, these
Caught in your hand to me.

Will you write in a new book
Just your name,
In a new book, for me?

LOVE AMONG THE RUINS

One time I was invited to see Croce.
Scylla and Charybdis lay between. In Sicily
The sky is blue over Taormina, flagrant
And dependable as a Greek statue.

The old columns were still taking the air.
It was a green, ethereal world, white Etna
Increasing the ponderable, inevitable nature
Of arcane spiritual adventure.

Over the sea Croce stared upon experience.
Insinuated within the discourse of events
A goddess appeared upon the stage declaiming.
She had ravishment, and sorcery, and authority.

Love was then the old debevilment
Of soul and sense; of fictions and mountains; of
Blue, green and of imperious whiteness.
Croce sat reading a book in Naples.

191

I swore to the gleaming gods of Hellas,
To the gloomy gods of America, and to these
Columnar shades above the turbulent waters
That I would search the white face of Etna.

I took my ease where the meanings meander
In crucial zones of mystery, where fire and ice
Commingle and life and death are one
And never went to see Croce in Italy.

ANIMA

We are betrayed by what is false. Within
Our hearts good and evil strive. Our minds,
Sometimes aloft, above grace, or sin,
Search out satisfaction. Time reminds

Our senses of the savage contradictions
Entertained tumultuously. We win,
We lose, we stay awhile upon convictions.
In some new century has any been?

The falsity is life itself. We are
Betrayed by time, which made us mortal. Time
Is a laughing light upon an ancient star.
Inmost thought is subtly made to rhyme.

It is the perdurable toughness of the soul
God and Nature make us want to keep;
The struggle of the part against the whole.
Each time we take a breath it must be deep.

THE SUPREME AUTHORITY OF THE IMAGINATION

Life longs to a perfection it never achieves.
Voices from the grave, with a wry grimace, whisper,
"We were never satisfied, we tried toward perfection,
We were undone. It is nothing to be a cipher."

Life breeds its joy in the incomplete.
A flag flying in a moment of the Atlantic
Lifts the spirit, lavishes a serene sign,
As halest happiness is death to the frantic.

Aesthetic purity stays with the rose bud.
Desire is renewed in every lover's glance,
The ununderstandable is always understood,
It is the shyest by the wall most wish to dance.

It is the grace to imagine the unimaginable
Elevates man to an angelic state
In which he may dwell, formidable and alone,
In the simplicity of the truly great.

It is the supreme authority of the imagination
Brings a frothed brightness to the human scene,
Bottle of industry, looking glossy-fine,
The blood-tip breezy and the anchor twined.

A rose of Spring seen with an even eye
Never betrayed the seer; he leaps the sight
And stands within ineluctable dominions,
Saved in some haven of a sheer delight.

PERCEPTION AS A GUIDED MISSILE

I

Resume unequal days, in chilly thrall
And in too hurting prisons; these four walls
For vistas plain; or those assumed through them,
My placid North, snow-bound in eyes as cold;
Upon my breast, with burn, the opposing South;
Cloy of ease, this East I nostril well;
And the windy gap, the damaging gray West.
But feet that feel are met by four plain walls.
His face! And all His grace habitual.

There is by snow a visible arrival.
I face this wall. Her face was dead, was clear.
Around, and by fire we are born intense,
Consume the essential gift desire, and feel.
I turn. Without giving or taking, lithely
Suspended, all timelessly is joy sweet.
I turn. And cringe at the brute savagery
Masterfully yelling, crazed by its own storm.
This; this; this; this; they seem the same,
Here prisoned where I count my steps all day,
In chilly thrall, and under the soft glare
Of the secret face of one adoration,
Attained; unattained; and unattainable
If walls were smashed, the sun plucked, and stars.

II

These distinctions chill me more. Proceeds
The unpleasant enterprise; heavy the footfall
That knows four ways, by rote, and those by walls.
Compelled; compelled. Inwaft, seductively,
The mellow rays of Eastern healing flowers
I fabric for a close worn cloth-of-gold,
So delicate, it is my skin's patine.
But the snake, that feeds on flowers, from my right side

Makes hiss with fatal passion, and in love
To the violent given force, I kiss, in joy,
In pain, the sting, annihilating the attraction.
Then, O terrible, the wolf ravenous
Rages ferocious, and I turn to the left,
Powerless, helpless, in repulsive fear
To be destroyed, swift fangs in the back.

Imprisoned. I seek the essential North with eyes.
Where flies around himself the albino crow.

The passage out will warm my brows apart.
The sun is of all these compact and stained.
Once held her spirit like a shape of sorrow;
The vibrant touch can give all death's load.

Often, most unsuspected, the dead face
Appears and brings; to make the present world.

III

I've been far off on seas, where compass light
Told North, South, East, and West, but night did not;
Nor blood, streaming in the wilderness
Of being, mute with ocean, and over-ocean.
All was roundness. The moveless turning ball
Allwhere, there; viewless, seen; and supreme.

Cold refusal. Pure walls of nothing,
Mortared well. Only the feet feel, walking.
Walls are often gates, and open vistas
Prisons. Each vexes with an angry stare.
I burn, to note the golden counterpart.
And tell myself aware by sudden feel.

You are surprised? But in no debt to cold.
I close my throat up with an humble pin.
And fret this brick-plot with an ardent heel.

IV

Along tender tangents, variant shocks contend;
Since between East and West a needle roams;
And not to the West entire must hearing turn;
And somewheres are to dwell, not delicate North.
Say away, say away these positive forming walls.
Bare intellection, willful absolutes?!
Too cold too long is betrayal too bare, too coy.
To fix an electric fluid in a square,
This terrible creative mirth and ring,
This piteous conditioned pain and cling,
This to-be-else, my tasted thwarting blood
Mortal. If they are, to see things as they are.
Ha, a man's just come in with an oilcan.

V

O lovely sun (like a singing boy),
With linger sweet, and marvel tone
Soft-coming too, be joy and woe.
And this my means: here on my given
Flesh, my flaxen bells of pleasure,
I play your warm resplendent shine.

VI

It is improbable, the faultless look.
Walls are harsh, and now my signs renew.
Scents; hisses; yells; and silences.
But love to one white adoration springs.
These doors, like tresses of a girl's dress,
Fling out, to let us deeper in, not out,
Where captive in the generous earth, a single
Spike runs from the sun to captive us;
And flaws freedom with the sweetest pierce.

196

VII

The way by thaw is looking-glass of sleet,
If senses can see, but mostly prospect sweet.
Smooth, warm, I think it ravishing since found.
The voyage is contorted by hot rain
And hail: only can spit-spit back in vain;
Yet noon's so sultry-lovely, time can beat.
No path is rescued betwixt hard stars of frost,
Where to fall foul, and there the world's lost,
On sharp cogs to you; the space between impounds.
At last your passage is extremed with ice;
It cracks, but snaps not its white device.
Endure, O mind, in circle that completes.

So smile; kiss; groan; all's on thin ice;
Save the volant crow, my master dear.
Save Him, who saves the world from sin.

BY THE STREAM

Every morning I wash the slate clean.
Each morning I sang, no such morning has been.
It is a springing, a green New England scene.

I wish to be deft in the stored morning air,
The past was locked in an immoderate lair.
Where there was stress, there would I dare.

By the living stream of language I
Threw down my lunch, dozing by an oak: try
As I could, I could not reduce the sky.

It was on a clear stream I teemed dreaming,
Prone and prevailing; I was that I seemed;
On my pied eyeball a blazing beamed.

197

On my noontime I, flushed and fashioned, fed.
To the earth, a gift of gloss, I gave my head.
From the stream I took the living dead,

The little newt, the yellow cowslip, water-shades
Loving their life in comprehensive glades,
I the many-bladed among the blades.

In this sweet, superior contention resting
My soul, a sailing bird on water, was nesting,
My green soul; it grew in the time's cresting,

When all of a sudden dark was coming on,
The basket empty, fragile, my musings gone,—
A glimpse of Love, and in the brake an unicorn.

WHAT GIVES

I shake an absolute around the world.
It is this: love never dies. It is alive.
This I tortured out in fifty years.

What others show, what I may know, before
Lung-blasting Death usurps my spiritual theme,
I do not know, but think will show the same.

As after years of rigorous spectacles
When evil mocked, and passion ruled and blazed,
And fear was lurking in a mullein eye,

There is the stealthy health of an hiatus,
Growth to a grandeur, strong and round.
Bring on blessings, Time, stay and sing.

Strength grows and throws around us holy love.
It is this I count on to the end of time.
Love is the end of knowledge, and sublime.

THE OAK

Some sway for long and then decline.
There are those, a very few,
Whose rings are golden, hard, and just,
Like a solid oak all through.

Each year builds on another truth,
A suffering, a joy increased in gold.
You cannot see, viewing the edifice,
Whether it is young or old.

Some have it in them to keep close
To nature, her mysterious part,
Seeming strange to be so natural,
Nature married to perfected art.

Flesh will ponder its dark blame.
Mind will never mate the true.
But the whole being will rejoice,
Like a solid oak all through.

IN THE GARDEN

Memory is a watery flower, when watered
Will transplant a scent from A to Z;
It is September. There are zephyrs truly.
A gold-banded bee in russet loam

Throws off the earth, his heavy cargo
At last aloft. You are the medium.
In you the subtle messages suspend
The alphabeta particles of gauze

Realization claiming permanence.
For these zephyrs, these sights and sounds,
That heavy bee, those indeterminate scents
Are not the world, but they are you.

You know them sitting in the shade
In old September, by the rich red banks.
The zephyrs are your mind, a probable
Stored world of melos appetite.

THE LOST CHILDREN

And so the river moves,
Impersonal and slow
Today. Let us go,
Afraid to own our loves,

To the place of ice and snow,
The brink where yesterday
Two children, held aglow,
Walked time away.

In a moment they were lost.
Ignorance and innocence
Pulled them under and tossed
Their souls up from sense.

Now, O stars, you hold
Their small souls, two
Children cold and old
In lissome Spring the true.

And so the river moves
Impersonal and slow.
We watch. Let us go,
And hold close our loves.

A COMMITMENT

I am committed
To the spirit that hovers over the graves.

All greatness flails.
Even evangelical Aristotle

Deploys his systems into mysticism,
Obfuscating his clarity of sight.

Plato goes with us on a picnic
As we mammock our hamburgers.

The logic of illogic,
The illogic of logic plays through time.

Poetic justice
Still is to get what we deserve.

I dream, here, now
Of the two-faced caprice of poetry,

Clarity of day,
Ambiguous necessities of night.

A spirit hovers
Over the stars, and in the heart.

Love is long
And art is good to stitch the time.

APPLE BUDS

Apple buds will never bloom
But to remind me of her room,
Impersonally proffering

201

Spring, when she was suffering.
She cannot take them in her hand
Again. I cannot understand
Her suffering, her suffering.
It is brutality to sing.

THROWING THE APPLE

(*Based on a painting by D. H. Lawrence*)

Adam and Eve sat in their garden.
The day was bright and fair, but Jehovah
Was looking over the garden as warden.
Would human softness ever harden?

Eve took a bite of the apple,
Concerned with the bright sun's dapple,
As shadows emerged in the woodland.
Adam's concern was to grasp, to grapple.

We then see Adam, unaccountably,
Stand up dark and fierce and shrill.
He aimed the apple at Jehovah
And flung it at him with all his will.

THE GARDEN GOD

Style is the water out of Homer,
   The way it flows. It never flows
     The same from the stone orifice
      Down into the garden pool.

Mark then that majestic head
   And fount of heavenly declension,
     The bubbles in the basin
      Making a merry noise and music.

Style is large and god-like, as if
    An elegance made life-like, and
        It seems to be an absolute,
            Curly locks, blind eyes, and sound.

Yet every day the water plays
    Differently from Homer's mouth.
        Each time I look thereon
            New enticement amazes me.

The winds of time will play the tune,
    A bird will come to sip alone,
        A cat will change the water-scene
            With arch and ritual concern.

I do not see our Homer plain.
    I see him melodious or restive,
        I feel the tensions of the day,
            The quick alarms, acclaims of change.

Style is magical despair
    Lest from the fount and source and power
        A single drop or grain
            Should lose itself from life's rich main

And it is plain for each to see
    In dream and consequence,
        The order of the oak and grass,
            The waving water's sound.

Or so I think replete
    With multitudinous harmonies
        On any day in summer here
            In the garden where our Homer is.

LIGHT FROM ABOVE

The vigor and majesty of the air,
Empurpled in October, in an afternoon

Of scudding clouds with sun breaking through,
Showing a militant light on mountain and river,

Is the imperial power
Greater than man's works

I praise and sing; my headlong delight
In unsymbolic gestures of eternity;

For here, surely, above the worn farms,
Their stoical souls and axe patience,

Whatever man learned from the soil, from
Society, and from his time-locked heart,

Is the greater, the grand, the impersonal gesture
And the imperial power; here, the great sky,

Full of profound adventure beyond man's losses,
Tosses the locks of a strong, abrasive radiance

From the beginning, and through the time of man,
And into the future beyond our love and wit,

And in the vigor and majesty of the air
I, empurpled, think on unity

Glimpsed in pure visual belief
When the sky expresses beyond our powers

The fiat of a great assurance.

AUSTERE POEM

Life in its brief episodes
Of sordor or of glory
Stains the flesh,
Baffles the intellect,
     Forces truth upon us.

Love in its vainglory,
Its deep commemorations,
Is at best a truce
Before subtle battles renew.
     Truth is forced upon us.

Hope renews egoism,
There are times of harmony,
Yet darkness obliterates light
And death overcomes life.
     Truth confounds us.

What is the best attitude
To take in incertitude?
What are pain and pleasure
But dazzled by enigma?
     May truth surround us.

Locked in mortal force
I speak in a locked estate.
I bend to my endeavour
And spend my love for you.
     May truth breed life.

HOOT OWLS

Owls that cry in the night,
(I have noted toward dawn)
Hooting in call and in answer,

Were supposed to be mysterious;
Indeed, their voice is enchanting,
With a far-away roll and fall,
Yet imagine my distrust
Of what used to be poetical
To learn that these dull birds,
Instead of leading us on
Into labyrinthian melancholy,
Actually make their hoots
To petrify, as it were,
And scare stiff small animals,
Who, thus mesmerized,
Cannot move for deadly fear,
Whereupon the hungry hoot owls
Wing down swiftly, killing
Their prey in handsome talons.

We must go elsewhere
Than the cry of the hoot owl,
Somewhere beyond realism,
For the tone of the poetical.

TREE SWALLOWS

One of the surprises of nature,
Sitting on an open porch
Overlooking oceanic summer waters,
Is the well-dressed, elegant
Tree swallow whose grace
Defies a sweep of pen
As he rounds a curve and,
Coming from out by the rocks,
From over the summer ocean
Drops down by your head,
Only an eye-shot above it,
Swiftly across to the tree
Ten sleek feet away,

Pulls up abruptly to it,
Places his feet on the rung,
And looks in, and goes in,
And then looks out adroitly
With a tailored, deft face,
Then steps out again
Eagerly, easily taking the air
In accomplished circuits
Catching gnats on the wing.

Especially toward evening
I watch his habitual flights,
Up high and aloft in air
Far out among the trees
Over the free coastline air,
Lost sometimes to sight,
But always coming back
High and swooping down,
Perfectly accoutred,
And as days go by in June
Every summer it is the same,
A svelte ritual of evening.
When he makes the nest
His mate will fly out and away,
And sometimes there seem
To be three, two males
And one female in the log-nest.
Elegance, grace and swiftness,
A well-tailored look, smoothness
And cheerful economy mark him;
As days and weeks pass
New duties come to the bird.

He takes his evening flight
As if in abandonment,
So easily, so splendidly
Swooping and turning and darting

But necessity is his
Like ours, it is purposeful,
He catches gnats on the wing,
Sometimes mouth-protruding bugs.
Now it is late in June.
He comes abruptly to the entry,
Just above my head and across,
I hear the little cries,
He stuffs each little throat
With a warm summer morsel
And off again, fair fatherhood,
To gather more for his family.
He seems carefree, seems eager
In airy freedom of motion,
Works around the apples and larches
And birches, up higher for mites
And returns, ever the elegant
Spirit of grace and duty.

In the heat of July
His capers increase, a pre-
Ordained change has come,
The parents go in and out freely,
The little ones peer quizzically
From the small round opening,
I dislike to see the days pass
Of such birdlinear charm,
But certainly as mid July
The last day will come,
Each day the same of each year,
The leave-taking. They are all flown.
And leave the mind to go
On its own accustomed flights
In its heavier manner
But bird-hearted all the same
For memory of leg-thrust
Up mountain rocks and leaps

Across chasms, and holding
The born clouds' lazy gauzes
As they twine off the peak,
And having an eye upward,
Being free, and coming home
To our own graces and duties.

THE CLAM DIGGERS
AND DIGGERS OF SEA WORMS

Appear far up the cove at low tide,
When the sea floor is a wet mastic,
Four men universal
In bent attitude of work,
Gray, mud-coloured, dun,
Caught in a moment of time
When a secret yield
Is possible to ancient earth—
I see them from a field
Under a cliff,
Almost static, scarcely moving
In their solemn grandeur,
The clam diggers,
The diggers of sea worms,
Placing their rakes down
Hard in the muck,
Loosening, pulling up,
Slowly manoeuvering,
Making rough black lines
In their slow progress
Disturbing the smooth, wet, black
Sea bottom of the cove
As sea animals may be
Making necessary tracks,
Solemn, ancient,
Then they became again
The living gray workers

Honest as surrounding cliffs,
Man making a living
From sea-fat residuums
When the sea has receded,
All gay boats and sails
Far off: earth-rich
Dark, gray men at work,
Impersonal as pines and sky,
I watch the heavy scene,
The slow, mute progress
Of torso, arm, leg and rake
As seeing a dark core
And sombre purpose of life,
Primitive simplicity,
Dignity beyond speech,
My mute salutation,
Time-deepened love
To clam diggers,
The diggers of sea worms.

A SHIP BURNING
AND A COMET ALL IN ONE DAY

When the tide was out
And the sea was quiet,
We hauled the boat to the edge,
On a fair day in August,
As who, all believing,
Would give decent burial
To the life of a used boat,
Not leave a corpse above ground.

And some, setting fires
On the old and broken deck,
Poured on the kerosene
With a stately quietude,
Measuring out departure,

And others brought libations
In red glasses to the sea's edge,
And all held one in hand.

Then the Captain arose
And poured spirit over the prow
And the sparks flew upward
And consigned her with fierce
Cry and fervent prayer
To immortal transubstantiation.
And the pure nature of air
Received her grace and charm.

And evening came on the sea
As the whole company
Sat upon the harsh rocks
Watching the tide come in
And take the last debris,
And when it became dark
A great comet appeared in the sky
With a star in its nether tail.

THE HARD STRUCTURE OF THE WORLD

Is made up of reservoirs,
Birds flying South, mailmen

Snow falling or rain falling,
Railmen, Howard Johnsons and airmen

Birds of Paradise
Silk lined caskets

Prize poems and guitars,
Beatitudes and bestiaries,

Children taught contemporary manners,
Time taking time away

With a haymaker or a sleigh,
Hope always belaboring despair.

Form is a jostle, a throstle,
Life a slice of sleight,

Indians are looking out from the
Cheekbones of Connecticut Yankees,

Poltergeists deploy northward
To tinderboxes in cupboards in Maine,

The last chock knocked, the vessel
Would not go down the Damariscotta

Until the sick captain's four-poster,
Moved to the window by four oldsters

Gave him a sight of her, and
He gave her a beautiful sign,

And there was the witch of Nobleboro
Who confounded the native farmers

Who, having lost the plow-bolt
Right at their feet, found it

Concealed in her apron: she laughed,
And made the earth fecund again.

The hard structure of the world,
The world structure of illusion.

From seeing too much of the world
We do not understand it.

There is something unknown in knowing.
Unfaith is what keeps faith going.

THE PARKER RIVER
(*To F.R.K.*)

I

It must have been a profound and a serene communion
When nearing dawn, still in pyjamas, to walk away
The body of this world, and give the body of yours
To the salt and running waters by the sea.

I gave my own son his first and baby dowse
In the Parker River: he cried with pain and joy together,
Knowing not that to go back to the sea
Is the great symbol of original mystery.

II

Is it right to take your life
Or better to let life take it?
Is it superiority of man
When things have come to such a pass
As his, the whole truth none else knowing,
To make a final affirmation
In a final, great negation?
Is it an intellectual grace
So to do, and, Oh far deeper,
Is it a spiritual unity
Known once and forever known
Between the doer and the Creator?

Or is it a cold, implacable knowledge
Of the lessening river in the blood,
A last defeat before a mask
Life always wears and never loses,

Inscrutable in its designs?
As who would say, I am the lost,
Forever lost, I am nothing now,
Of nothing, neither bad nor good,
My action is my final statement,
I too will be inscrutable
And none shall know my final meaning.
Mysterious action of a man

To walk into the river waters
Leaving wife, sons, time behind him,
The tidal river of the marshes
Sinuous and ancient, seaward
And landward ever moving,
As years move forward to our future
And backward to our past remembered,
Or caught in stillness at full tide
When time seems not, is not insistent,
And then the river seems a gauze
And we are perfect in its laces
Beyond action, within contemplation.

III

MOON

I incline and I decline,
I long toward mankind,
Oh in my passive glow
I all enchantment show,
Then, growing faithless-glimmed
To erring man who never
Loves me beyond my changes
And in his wild heart always ranges,
As if my own perfections were
Instead of brilliance, a dim blur,
I yearn back to father night
And pass mankind from my sight.

WIND

I try with every fret of feet
In measure like a dancer, neat
And true to every restlessness
And every hint of evanescence,
In my brunt and airy measure
I give with love my treasure
Rapt or easy, swift or slow,
Over field and meadow go
And come upon the human brow,
The shoulder, the sensing torso
To tell the secrets of the earth,
Coming from the realm of birth.

STARS

I am the spirit of the stars,
The all star, one welling impetus
August and final, beyond humanity,
Beyond wisdom, beyond finality.
And thus I am perfectly wordless
Serene beyond anything heard,
To whom the trials of mankind
Are like water running over sand,
So obvious and so mysterious
As to be beyond all seriousness
And in my timeless innocence
I contain all of experience.

NIGHT

But when weariness and harshness
Bring mankind into darkness,
When he pales upon the light,
Lost in the hives of sight,
When the strong are heartbroken
And in the lists of life no token,
Not one is worth the holding,

Such great mystery unfolding,
I am the dream of sweet peace
And the angel of an ease
Like water, like mother water flowing
Beyond the effort of man's knowing.

DAWN

And I am the dawn beyond life,
I come inevitably as the sea's return
In the small and happy estuary,
After the melancholy blankness,
With slow steps, and solemn renewal
Where the birds awaken to the true,
The most slow and certain beginning,
The inevitable, quiet winning,
I am the pure solution,
The forgiveness and the resolution,
The aspiration, the hope, the new breath,
I am the glory beyond death.

EARTH SPIRITS

We sing together our ancient song
Over man living short or long,
Beyond his breath, beyond his will,
It is strange kinship still
Whatever happens from his birth
In his pilgrimage on earth,
We too are mysteriously held
And mysteriously impelled
And in one nature we exist
And through one evidence persist
As river waters are bridal and coronal,
Baptism and eternal funeral.

IV

Sons of his and progeny of mine
May swim again in the Parker River,
Loll in breast-deep fullness of the tide
Or let the current take them out, or inland.

There is a bridge near the swimming place
To go under, one way or another,
Alone or together, knowing or unknowing,
Loving the eternal flowing water.

AT THE CANOE CLUB

(*To Wallace Stevens*)

Just a short time ago I sat with him,
Our arms were big, the heat was on,
A glass in hand was worth all tradition.

Outside the summer porch the viable river
Defied the murmurations of guile-subtle
Truths, when arms were bare, when heat **was on,**

Perceptible as picture: no canoe was seen.
Such talk, and such fine summer ease,
Our heart-life against time's king backdrop,

Makes truth the best perplexity of all,
A jaunty tone, a task of banter, rills
In mind, an opulence agreed upon,

Just so the time, bare-armed and sultry,
Suspend its victims in illusion's colours,
And subtle rapture of a postponed power.

OSPREYS IN CRY

When I heard the call of the osprey,
The wild cries of the ospreys
Breasting the wind high above
The cliff, held static
On updraft over the ocean,
Piercing with ancient, piercing eyes
The far ocean deep

I felt a fleshed exultance
For the fierce, untamed beauty
Of these sea-birds, sea-hawks,
Wild creatures of the air,
Magnificent riders
Of the wind's crests, plummeters
Straight down for prey

Caught under water in talons
Triumphant as life,
The huge birds struggling up
Shaking heavy water off
And powerfully taking the air
With fish in talons head first;

I felt a staggering sense
Of the victor and of the doomed,
Of being one and the other,
Of being both at one time,
I was the seer
And I was revealed.

HALF-BENT MAN

Haunts me the lugubrious shape
Of a half-blind, burly, old
Man, half bent to earth

Who on the Princeton campus
Spears stray papers with a nail-
Ended stick, lurching, walking
Crankily, true scavenger
Ridding the earth of detail
And debris, always evident,
A bent man, the true condition,
Half-blind, but cleansing life,
Putting trash in a burlap bag,
He moves as it seems to me
With a profound, heavy purpose,
I am haunted by his life
As towers, books, professors, ideas
Mingle in a world beyond him,
But it is his own, dark burdens
In his bent, half-seeing, weary attitude
I claim as man's and mine,
And O blind-man, rag-picker,
Paper-picker, cleanser of domains,
I shall not betray your meaning
As time bends us to the earth
And we pick what gems and scraps
There are from magnificence.

SPRING MOUNTAIN CLIMB

Till thinking had worn out my enterprise,
I felt, and felt the flesh
Salt-swart, blood-sweet,
To which bird-song stung mysterious
And the white trillium mysterious in the wood;

I saw the mountain and the lake,
I followed where the source sounded,
Over boulders, crossed logs,
Up rugged reaches, where
The gates of mystery increased;

Sacred justice moved me, I entered
The ancient halls of visionary grace,
Bird-call sounded, sky
Appeared miraculous along,
And evil thickets held red histories;

Here divine justice made me sweat;
While an eye nebulous and profound
Partook of a huge nature, endless
Sufferings redeemed in rushwater,
Song falling, searching the world.

Such were the signatures I saw
Written by the hand of God
In knotted density, mysteries
Of incontestable day, while I
Passed the singing brook, so

Controlled by its eternal sound
As to be a living witness
To spirit, and to spirit reaching,
And to the sound falling back,
And man fallen to his endless burden.

THE PASSAGE

Disindividuating Chaos
And old Discord clamped

Down on my downy love
Before it was spoken of,

Suckled must be in a year
First fingered; sensed no fear,

Then shot up in the blue sky
Conclamant with ability,

O I remember the holy day
When glory along me lay

In brightest shoots, singing sunbursts
And honey-great thirsts.

Then power came with outer throng
And dense strife of tongue.

II

With power came delight
Which put the world to right

Before ever it was wrong,
Incredible joy, poem-song.

Early then I knew
It was a gift of the true,

I struck a dangerous course,
Reckless a spendthrift. Source

Was sure, great world-brightness
Washed clean in lightness

Opened along my ego,
I danced alone. I would go free

With a will never tame
In the soul's single aim.

III

I took the world alive,
Drew honey from the hive.

Experienced quickened,
Then discourse thickened,

The heaviness of a fall
Fell like a blight over all,

I jostled in sea-spate
Of world-wrestle, dared fate

To tell me the worst
And fell on pain, and cursed

The doom upon the race,
The death in every face,

And when I saw men die
I heard their holy cry.

IV

Dense was word-drift,
Nor could the senses lift

To purity of Psyche dream
But dream would deeper seem

With knowledge in every breath
Of loss, suffering, death.

I came into dark hours
In the loss of powers

And wove my life with men's
Detentive stratagems.

I spoke beyond the nation
In imagination

And loved the mortal mind
Of timed humankind.

V

Language became
The unifying aim

In density and purity
Of all we can see,

The statements of the eye
Gained strong clarity,

The reaches of desire
Became a holy fire,

Words of fire to fashion
The song of man's passion

And over the brooding hurl
Of world's meaning, a furl

Of peace and ease, sight
Of unity, the holy light.

VI

But ambivalence and terror
Struck everywhere in error

And in errors of living
My world-force was giving

Compulsions of the blood
In pluralistic love

As effort steepened,
As years deepened,

Darkened to satiety
By a palled society

Within materialism,
Without aerialism

Of the soul's light
And joyful fight.

VII

Beside a river I stood
Nearby in a wood

Where was a spring
And in it a thing

That looked like a cup.
I wanted to lift it up.

It was hidden in mould,
Looked broken and old.

I bent, and reached down
To an invisible town

Where antique lovers danced,
In bright air entranced,

And came to grips,
And put it to my lips.

THE GODS OF WASHINGTON, D.C.

I was wondering about the gods
As I was walking down Constitution Avenue
For always since I was a boy
I was conscious of the gods in the offing,

Always off there somewhere in the air,
Never immediate and satisfactory,
Yet indubitably the master stuff
And living and seeing of our being,

And now here on Constitution Avenue,
With memorial buildings austere and white,
I wondered about the gods of the world,
Their existence, their glory, and their authority.

The gods were so plural and so vague,
Yet so powerful and so composite
That I was perplexed on Constitution Avenue
Even as I had been perplexed as a boy

And nurture this poem about the gods,
Who spring and leap out of our blood,
As one still seeking for definition
At odds with, but admitting the gods

And if they were a fallacy or illusion
May they go back before their birth,
But if they are a waking fact and a conclusion
May they state their case and return

For in this century of delicate balance
We have need of the gods of the air,
Stout gods, stubborn as bony man,
But we ought to know whether they are there.

EQUIVALENCE OF GNATS AND MICE

As a pillar of gnats, moving up and down
In June air, toward opulent sunset,
Weaving themselves in and out, up and down,

As diaphanous as visual belief,
In scintillant imagination, is slightest dancing,
Weaves a major harmony of nature;

As tiny field mice are saved from the sickle
By a lean seventy-year-old scyther in Maine
Who brings them in, saying, "They have enemies enough";

Who are hand-fed by a dropper on milk and water,
Hoping the small creatures will survive and thrive,
Slight event against the history of justice,

It is necessary to hail delicacy
Whenever encountered in nature or man;
No disharmony come near this poem.

BIRTH AND DEATH

I dropped to depth,
And then I leaped to height,
But in between was the fearsome place.

All imaginative skill
Could not shape the in between,
While depth and height were absolute.

I called in philosophy,
With mortarboard on top, to help;
The serious gambit of wisdom.

And I called on love,
Great bounce, good thrall,
Knowing the happiness of lovers.

And action called on me,
Saying, take the sea, the sky;
Yours is the enthusiasm.

It was like the drama
Of a man thinking he
Could outwit life itself,

Like the search for purity,
Irrational struggle of the will,
A might and power and force.

The in between will not be conquered
So long as man shall strive
And struggle with the world.

I dropped to depth,
And then I leaped to height,
But in between was the fearsome place.

THE INCOMPARABLE LIGHT

The light beyond compare is the light I saw.
I saw it on the mountain tops, the light
Beyond compare. I saw it in childhood too.
I glimpsed it in the turbulence of growing up.
I saw it in the meshes of meaning of women.
I saw it in political action, and I saw
The light beyond compare in sundry deaths.

Elusive element, final mystery,
The light beyond compare has been my visitant,

Some sort of angel sometimes at my shoulder,
A beckoning guide, elusive nevertheless,
Under the mind where currents of being are running,
It is this strange light I come back to,
Agent of truth, protean, a radical of time.

The light beyond compare is my meaning,
It is the secret source of my beginning,
Issuance of uniqueness, signal upon suffering,
It is the wordless bond of all endings,
It is the subtle flash that tells our song,
Inescapable brotherhood of the living,
Our mystery of time, the only hopeful light.

MAIS L'AMOUR INFINI ME MONTERA DANS L'ÂME

I

Her eyes are like the putrid sores of infants
Born into the world from syphilitic mothers,
Her mouth is a joke, a leer, a sewer,
Her teeth lacerate her tongue when she speaks.
  If my mistress is different than this
  Discover it to me, come show me.

We will look lower, and we will see her breasts
Like a musty barn full of rotten fodder,
Like two young rats, consumptive and affectionate,
Gnawing the daylight in a purple fever.
  If my mistress is different than this
  Discover it to me, come show me.

But we will look lower still, where she bleeds.
Where God should have split her, He made a question mark.
And her soul shall twitch among the snakes
Altogether in a nice, yes, a nice way.
  If my mistress is different than this
  Discover it to me, come show me.

II

I saw my mistress change into an animal
By the grace of God, by the Devil's hard decree,
I saw her face coarsen and get hair upon it;
Her superb form reduced to a laughing jackal's.
  When my mistress is big with human love
  Bring her to me, proffer me that offspring.

Her voice, that sang the lusty praise of heaven
Now could only squeak, splitting the air.
Her lovely motion, kindly, graceful and sedate,
Shattered itself into vicious kicks and jerks.
   When my mistress is big with human love
   Bring her to me, proffer me that offspring.

Her endearing eyes, that understood compassion,
Her breasts, that knew to heal man's thirsting woe,
Her hands, that kept the poise of simple duty,
Are fierce now, or between her legs, or hard claws.
   When my mistress is big with human love
   Bring her to me, proffer me that offspring.

III

As I sat by the evening sea, listening,
Up from the waters came Mental Progeny,
An army of men they were, prognosticating,
And at their head was the Poet, bringing Poetry.
   If my vision is false, blame the sea.
   Drown me under the waves of natural birth.

I saw them marching in order, calm and tried.
They had come to quell man's wasteful rage.
Before they should die, and their eyes would be holes,
One saw in their eyes comprehension and serenity.
   If my vision is false, blame the sea.
   Drown me under the waves of natural birth.

And the sea bore up a woman who was Love.
With eventful flesh, elate, she seemed to smile.
There the poet on a scroll wrote a psalm,
Then for the time being mankind was redeemed.
   If my vision is false, blame the sea.
   Drown me under the waves of natural birth.

ON SEEING AN EGYPTIAN MUMMY
IN BERLIN, 1932

Pain is seen to have been transitory
However severely it wrenched the twisted form,
It is seen twisted by more evil time,
Peace unto it, death to it, and the ages.

Joy is known to have been wholly false
However famously it waxed the cheeks.
Cursed be that which estranges us most,
It is consciousness blights us all.

From the ravages of pain and ecstasy
Deliver me, my human will, if you can
Till like an embryo, and like a mummy
I am curled in peace, like you, O happy Egyptian.

THE SPIDER

I

The spider expects the cold of winter.
When the shadows fall in long Autumn
He congeals in a nest of paper, prepares
The least and minimal existence,
Obedient to nature. No other course
Is his; no other availed him when
In high summer he spun and furled
The gaudy catches. I am that spider,
Caught in nature, summer and winter.
You are the symbol of the seasons too.

II

Now to expatiate and temporize
This artful brag. I never saw so quieting
A sight as the dawn, dew-clenched foot-

Wide web hung on summer barn-eaves, spangled.
It moves to zephyrs that is tough as steel.
I never saw so finely-legged a creature
Walk so accurate a stretch as he,
Proud, capable, patient, confident.
To the eye he gave close penetration
Into real myth, the myth of you, of me.

III

Yet, by moving eyesight off from this
There is another dimension. Near the barn,
Down meadow to shingle, no place for spiders,
The sea in large blue breathes in brainstorm tides,
Pirates itself away to ancient Spain,
Pirouettes past Purgatory to Paradise.
Do I feed deeper on a spider,
A close-hauled view upon windless meaning,
Or deeper a day or dance or doom bestride
On ocean's long reach, on parables of God?

SEA-RUCK

Washback of the waters, swirl of time,
Flashback of time, swirl of the waters,

Loll and stroke, loll and stroke,

The world remade, the world broken,
Knocked rhythm, make of the slime,

The surge and control, stroke of the time,
Heartbreak healing in the grime, and groaner

Holding its power, holding the hurl,
Loll and hurl, power to gain and destroy,

The tall destruction not to undo

A saffron inevitable sun, far and near,
Some vast control, beyond tear and fear,

Where the blood flows, and nights go,
Man in his makeshift, there is home,

And the dark swells, the everlasting toll,
And being like this sea, the unrolling scroll,

   Stroke and loll, loll and stroke, stroke, loll

THE HAMLET FATHER

When Hamlet had sunk to the moist ground
With his most meanings tossed
Back to the unwilling, pregnant sky,
His will and green questions lost,

I thought I had outlived his mark,
Viewing him with thanks and saying,
Hamlet, you are too young to count,
I assign you to the philosophical dark.

Mine are less modest and princely
Lucubrations on the same events.
I am your father and would care
To have richer evidence.

But had you lived longer and deeper
You might have gained a passion
Profound as the master of yours
And lived in a different fashion.

FOUR EXPOSURES

1. Playful birches in the
            Where the mind goes off in austerity
                Taking the baby to the doctor
      :Flexions
      Impressions:
      The late quartet draws the heart out
           Cash the check in the morning
                Polio may be latent for years
                Improvisations:
                Destinies
      The sun kills as it blesses
           King Canute knew the tides' tactics
                Mallarmé was foolish in his lectures?
                :Seizures,
                Creations:
      Never to be serious! Let the jest
           Be on one. And no more philosophy,
                Know more. Will Star's pups be thorough-
      bred, or mongrel?
                :Plumbers,
                Carpenters:
      How the world runs to worldliness.
           Strength of the soul! Passionate devotion!
              Go and get honey rolls and streptomycin.
                  Hush-a-bye
                  Hush-a-bye

2. Rich in dextrins, maltose and dextrose
           And nobody knows the answers. Here
                Comes Aurelia, with the baby carriage
      :Dissemble
      Invade:

Once having learned Chinese their subtlety
        In linguistic inspissates our English.
                Indian music babyfies J. Bach.
:Pretend, and
Invent:
Twice Washbowl our Cape cat had kittens
        In the same closet. At four of the morning
                She mews, and I let her in the window
                        :Salutations
                        Historicities:
Shall I allude to Mrs. Phineas Cuff?
        Everybody has a mad neighbor.
                Keep your dog from hence, we have a mummy
                        :Bosch
                        Angelus Silesius:
They have no bock beer this year.
        Pale genius is burning the midnight oil
                The literary politicians are giving out the prizes
                        :Petrifactions
                        Liquefy!

3. I dickered with the truth, because
                The truth dickered with me. Fractious
                        Divagations please the senses
        :Presentiments
Hallucinations:
Old years laugh hysterically
        At the seriousness of the new.
                We have seen too much of death in our time.
        :Calculations,
Fervors:
If of anything we were intellectually certain
        We could not afford the luxury of the pit.
                We would be lost in the found.
                        :vastness,
                        Unfound me:

The source is newness evermore uncreate!
It is delicious hardship to steal out of it!
It is handsome to handle the eventuating.
:Gusto
Diamonds:
The baby with her goddess–descending smile then.
Why not? Others have felt it before.
This assurance beyond us from the skies.
:Taken up
Defended:

4. The worst then ever the best. No defeat
That is not a subtle victory.
No victory that is not a defeat.
:Accept
Listen:
It always disturbed me that Job
Gave up his boils. I see now that tragedy
Is good only in a certain framework.
Shakespeare stops.
Homer goes on.
Birds on the boughs, lovers in the haystacks,
Cadillacs, Fords, and cold–water flats.
Lovers couple the theory and the fact.
:Vigor
fashion:
Time is the soul's macaronic specialist.
As the wheel turns, what will come of the wheel?
What songs, what waters, as the wheels turn?
:Trial
Aptitude:
I should have thought the senses best.
Let Plato, let Aristotle lie.
Lie to me then in the white night
Lover
and bride.

LA CROSSE AT NINETY MILES AN HOUR

Better to be the rock above the river,
The bluff, brown and age-old sandstone,
Than the broad river winding to the Gulf.

The river looks like world reality
And has the serenity of wide and open things.
It is a river of even ice today.

Winter men in square cold huts have cut
Round holes to fish through: I saw it as a boy.
They have a will to tamper with the river.

Up on the high bluffs nothing but spirit!
It is there I would be, where an Indian scout was
Long ago, now purely imaginary.

It is a useless and heaven-depended place,
Commodious rock to lock the spirit in,
Where it gazes on the river and the land.

Better to be rock-like than river-like;
Water is a symbol will wear us all away.
Rock comes to the same end, more slowly so.

Rock is the wish of the spirit, heavy symbol,
Something to hold to beyond worldly use.
I feel it in my bones, kinship with vision,

And on the brown bluffs above the Mississippi
In the land of my deepest, earliest memories,
Rushing along at ninety miles an hour,

I feel the old elation of the imagination.
Strong talk of the river and the rock.
Small division between the world and spirit.

LOSS
(*To V. R. Lang*)

Her loss is as something beautiful in air,
The mysterious part of personality
Become the blue mystery of the air,
The far and the near.

In life she had laughter and acting.
She made things gay and severe.
The world continues, beyond reason to fashion,
The far and the near.

She took many parts, she had only one,
One was her sureness of being.
The others were maskings of dark and light,
Her feminine grace of seeing.

I do not know how to say no
To time that goes in any case,
Do not know how to explain
The pure loss and vision of her face.

TO AUDEN ON HIS FIFTIETH

Dear Whizz, I remember you at St. Mark's in '39,
Slender, efficient, in slippers, somewhat benign,

Benzedrine taker, but mostly Rampant Mind
Examining the boys with scalpel and tine.

I recall the long talk and the poems-show,
Letters sprinkling through the air all day,

Then you went down and put on Berlioz,
Vastly resonant, full of braggadocio.

I look at your picture, that time, that place,
You had come to defend in the American scene

The idea of something new; you had the odd face
For it, books sprawling on the floor for tea.

It was the time of the *Musée des Beaux Arts*,
Your quick studies of Voltaire and of Melville,

In the rumble seat of my old green Pontiac
I scared you careening through to Concord.

And one time at a dinner party, Auden,
You wolfed your meal before the others were served

Treating the guests to an Intellectual Feast
Probably better than any of us deserved.

I remember your candor and your sympathy,
Your understanding, your readiness, your aliveness,

Your stubby fingers like lightning down the pages;
Our ensuing American years that made you thrive.

Now you are back at Oxford, an Oxford don,
Half a century gone into the Abyss of Meaning.

Here's my well-wish on your fiftieth,
You flex a new twist to the spirit's feigning.

TO WILLIAM CARLOS WILLIAMS

I would make this all as single as a song,
My own assumption in a flittering stance,
Twenty years cast in an easy affirmation.

The truth is there is truth on every side,
Each protagonist as relativist
Invests the present with his intellectual twist.

You are no absolute, Bill! But genial soul
And spanking eye, no hatred of your fellows,
Concludes we love you the worldly American.

With gusto to toss the classics out, and with them
The sonnet, you live yet in a classic Now,
Pretend to advance order in your plain music,

And even preach that Form (you call it measure,
Or idiom) is all, albeit your form would mate
The sprawling forms, inchoate, of our civilization.

NEXUS

The dead are hovering on the air,
So real they have their flesh and bones.
They appear as they had been,
And speak with firm, daytime tones.

I say, I cannot believe your power.
Go back into the ancient times.
The sun burns on my forehead now,
And thought comes in a spring of rhymes.

My love is like the blue of the air,
My son and daughter play at games.
We live in a yoked immediacy,
Imagination come, that no one tames.

Everything I do today
Moves with a stealthy, spirit strength,
A thrust into the future order,
But yet it has a backward length.

The dead are playing about my head
As real as present, effable air.
They have their power to make and shape
Each breath I take, each thought today.

EXAMINATION OF PSYCHE:
THOUGHTS OF HOME

Now at ninety this frail and lightsome woman
Goes to the hospital to meet her death.
The long decades stretch as a panorama
And I think of a slow third of a century
She has lived beyond the one I loved.
The one I loved has long since gone to dissolution.
The bare perplexity astounds me still,
How all that love she bore me, and I returned
Fullfold in fond, filial devotion
Saw but shadowy reflection in this other.
Yet beautiful she is, always a reminder,
A triumphant form, statement of life's harmony.
Now at last this delicate-boned, fair woman
Goes toward her death and makes one ponder
On fortune's intricate ways; childless she goes,
Who lived out almost a full century,

In one town, among certainties and changes.
When the mind is ravaged by thought
There is no virtue in the passage of time.

Is there a new myth to people the mind's eye?
Is there a splendour undreamed of by man?

I ask these questions in the name of love
Bestriding life. I think of a delicate-boned woman
In her long, harmonious relationships
Which put off answers to intolerable questions
Assailing me with that other, her friend,
Long lost in the wry, unanswerable enigmas,
And in the name of spirit, likewise bound
On a quest for an ultimate answer,
Here in mid-dream, big and hardy, but alone,
Let me make an epiphany for the living,
For those who die early, for those who die late.

THE PROJECT

The mail box, the roadway, and the dump truck
    a day in June blue and gold

The play rock, the high swing, the pine stand
    a race of children readying time

A prospect of a legendary Connecticut
    angler flicking his fly in amber evening

Tumble of waters at Diana's pool
    Dive, shimmer and roil of bathers, devotees.

It is the pageant of the American summer
    Blue, gold, and high, an elegance of time.

When freefoot lads find a decaying animal
    In the woods, one yells "Help! Don't breathe it!"

MATADOR

It is because of the savage mystery
There in the coffin, heaved on burly shoulders,
At five o'clock in an afternoon of jostling sunlight,
We wake to the rich meaning of necessity

Close to the horns, on the horns of the dilemma
Instantly tossed, gored by the savage animal,
The dance in the bullring flares sense magnified,
And turned and tended to the pains of perfection.

Matador of the spirit, be you also proud and defiant
By grace and skill, accost hot sunlight without fear,
Try nearer to the fetish tossing of the horns,
Relaxed power best defies the brutal adversary.

And hold that skill most dear that most dares,
The dance almost motionless, as the beast passes,
At five o'clock in an afternoon of jostling sunlight,
Were crowds, and banners, wilderness, and music.

PROMETHEUS

Touched by fire,
Holding the light in my hand,
Brilliant-colored October,
Adoration is the only word.

Breathless I am,
I hold the light in my hand,
The red, the green, the tawny,
Blue sky illimitable,
Adoration is the only word.

243

The light is on my hand,
Changeless golden afternoon,
Gift of the metaphysical,
Deep infusion of the seen,
Adoration is the only word.

And adoration is the only word
As, holding the light in my hand,
The light is on my hand,
Striking dumb,
Touched by fire.

OLD TOM

An old, black, rutting tomcat,
The brother of his female,
Expressed nature in his sister
Begetting again his future.

I eye this old, mangy fellow
With a certain sympathy.
His progeny already
Have suffered fortune and misfortune

Teaching us, as larger animals,
Something of ourselves.
As poets will to survive,
Cats survive by force.

A kitten could not be expected
To understand a moving car.
One, atop my front wheel,
Was rolled down to mutilation.

I had to kill her with a club
And buried her in the bushes,
Shaking with the dread of this
But doing it nevertheless.

Her little brother very soon
Had caught a bird so beautiful
I hated to see it mutilated,
And left only feathers and the spleen.

They are the most civilized creatures,
Sleep all day and hunt by night,
Elegance in the drawing room,
Merciless in dusk or in moonlight.

But most it is their indifference
To death of their own fellows
I applaud; they go about their business,
Unquestioning the fates of those fellows.

Old Tom, here is a handout,
Some meal and some milk for you.
Go rough it under the stars,
You teach us what we are

When our policies are riven
And our pretentions are bare,
And we are subservient to nature
Very much as you are.

THE HEIGHT OF MAN

I am where the bluebell dies
But I, where Western storms are born,
Am ready for new mysteries,
I think of riding the bullock's horns

To sweaty dust; I climb Hermit's Peak
Ten thousand feet up beyond trees
Where in the last reaches high crosses
Mark the stages of the Penitentes.

Some say their savage mysteries
Were so guarded on the heights
That when a non-believer climbed to them
They slew him with a light slingshot.

Bluebells, bluebells, how frail and Spring-like,
Another world all fragile and intact,
O bluebells of the early memories and life,
A fair elixir before the killing facts.

AN EVALUATION UNDER A PINE TREE, LYING ON PINE NEEDLES

When I wrote the abstract of Heaven
I felt worse, and went to a pine tree
Where I studied its fallen needles
To see the pattern of the universe.

When I had made love to woman
I felt moral, as having conquered this,
And kept faith with a mystical union
Which is the saving grace of the world.

When I begat children, I knew I had lived
And surrendered my fullness to the future.
But always there was a spirit in the flesh
That the flesh demanded greater sacrifice.

I hung on godhead all a strained life,
Wrestling with notions of the supernatural,
Craving the finite taste of infinite essence,
Believing in the glory of disbelief.

I came back to the wonderment of nature
Under a pine tree, my skin abraded by needles,
Affirming the great distillate of joy,
Walking in the woods, inevitable, evanescent.

When I was fifty years old, I felt
The grandeur of my volatile ignorance
And I made words as ruthless as an arrow
To pierce your heart with vigour and resonance.

Therefore let us rejoice, and though I offend
And there is no end of suffering,
Accept my song, be rich in my nature,
Bend to the northwind, and to the pine song.

KAIRE

If I were Sophocles, brave with truth,
Writhing in the darkness of humanity,
Bright with an occluded brightness,
Able to hold in total mind
The fantastic reality of the human condition,
I wonder what I would have done
About a world beyond the Greek,

For he knew the zenith and nadir
Of passion, and he knew that beyond reality
Was the other passion of mythology,
That myths were sensual as tears or dreams,
The stains of error in the habit of truth,
Leaning beyond the flesh to the strength of the gods.

If I were Sophocles, ebullient and melancholy
Today, I would be unable to say
How far distant is the dream of Eros,
How divorced from primal concern seems truth,
How love is the power we but dimly see,
Love is that wholeness of the passionate mind
Glimpsed in the sensitivity of being;
The blind in our day see more than the seeing;
This is as the vessel in the enriching breeze
Knowing only immediately where the wind is blowing,
Yet time will take it to the mark
Eventually. Hear the lark
In its cry at dawn. Hear the stating
Breakers: before destruction they preserve you
To dream on a world of immortality.

A NEW ENGLAND BACHELOR

My death was arranged by special plans in Heaven
And only occasioned comment by ten persons in Adams,
    Massachusetts.
The best thing ever said about me
Was that I was deft at specifying trump.
I was killed by my father
And married to my mother
But born too early to know what happened to me,
And as I was an only child
I erected selfishness into a personal religion,

Sat thinking forty years saying nothing.
I observed all. I loved to drink gin,
Would not have thought to go farther
Into arcane episodes of the heavier drugs,
And, being New England, always remained sober.
However, I confess now, I was
Always afraid of women,
I don't know why, it was just the way it was,
I could never get very close to any woman.
Knowledge and intelligence allowed me
The grand rationalization of this; also, I respected
Delicacy, but would not go too far in any direction.
I thought I was a good man. I was.

I did not obstruct the state, nor religion,
But I saw through both and maintained my independence.
I kept my counsels among the learned.
My learning was more private and precious than worldly.
The world had no sense of the devious,
So my private vicissitudes were mine alone.

I say all this with a special sort of grace
For I avoided many of the pitfalls of fallen man
And while I did not have heroic size, the
Creative grandeur, or mastership of the mind
I earned my bread by cynicism alone,
And blow you all a kiss from the tomb.

A MAINE ROUSTABOUT

He was there as the yachts went by
Percy is my name; my accent is good,
I am told, as good as that of an Elizabethan.
I had no schooling beyond the age of sixteen.
My wife left me. I took to drink, live with a dog.

I resent children unless they can hold their own
With grown-ups. I've been around the world on ships,
Down Connecticut way on jobs, once got to Georgia,
Always return to the rocks and the hard times
Of Maine. At clambakes in the summertime
I sit with the summer folk on the conglomerate shore,
Play my old fiddle a sharp tune or two,
Old airs I learned from my brother when we were boys.
It was always tough with me. Sharp as the city folks
I think I am, but am ever wary against them,
Keep my difference, and will not let them tell me off.
I have no respect for their savage villainies,
Yet their power over life always fascinated me.
They own the place. They come and go, I'm left
To chores and dung. But I can catch a mackerel
Almost any afternoon on the incoming tide
With an old hook, when they're running, old line,
In my old boat: they won't take hook from the richlings.
If I scare the children with my grizzled face
It's an old gut forced with whiskey keeps me going.

SEA BURIAL FROM THE CRUISER *REVE*

She is now water and air,
Who was earth and fire.

Reve we throttled down
Between Blake's Point and Western Isle,

Then, oh, then, at the last hour,
The first hour of her new inheritance,

We strewed her ashes over the waters,
We gave her the bright sinking

Of unimaginable aftermaths,
We followed her dispersed spirit

As children with a careless flick of wrist
Cast on the surface of the sea

New–cut flowers. Deeper down,
In the heavy blue of the water,

Slowly the white mass of her reduced bones
Waved, as a flag, from the enclosing depths.

She is now water and air,
Who was earth and fire.

FLUX

The old Penobscot Indian
Sells me a pair of moccasins
That stain my feet yellow.

The gods of this world
Have taken the daughter of my neighbor,
Who died this day of encephalitis.

The absentee landlord has taken over Tree Island
Where one now hesitates to go for picnics,
Off the wide beach to see Fiddle Head.

The fogs are as unpredictable as the winds.
The next generation comes surely on,
Their nonchalance baffles my intelligence.

Some are gone for folly, some by mischance,
Cruelty broods over the inexpressible,
The inexorable is ever believable.

The boy, in his first hour on his motorbike,
Met death in a head-on collision.
His dog stood silent by the young corpse.

Last week, the sea farmer off Stonington
Was tripped in the wake of a cruiser.
He went down in the cold waters of the summer.

Life is stranger than any of us expected,
There is a somber, imponderable fate.
Enigma rules, and the heart has no certainty.

RUBY DAGGETT

She, a woman of abrupt features,
Cocked an eye this way and that. Another
Decade went by. She sat still in her place,
Looking out at the alley from the bookstore.
She was the mistress of rich indices.
Who came, who went, who was in, who was out
She noted, impersonal above the day's charges.
Chocorua jumped from the cliff, the cash register struck.
Another decade passed, she sat erect.
Before she knew it another would come and go.
She, a woman of abrupt features,
In a small space, looking out at the alley,
Encyclopedic among files, ensconced,
Let vision flow over the peaks of time
As over the mountains of imaginative reality,

Watching the people and books come and go.
Ledyard cut his canoe, carved an archipelago.
She was like a peak. She was distant,
Who was always present. She was stalwart,
Like mountain silence. And she knew,
Take care of the day and the day will take care of you.
Roosevelt rolled up a ramp at Marlborough.
Ike beamed. Kennedy rolled up his sleeves on the green.
She knew the scent of books, as intimate as age.
In ink she suspected a waft of sage.

HARDENING INTO PRINT

To catch the meaning out of the air
Yet have it inviolably there,
Life I mean, the glimpse of power, incomparable times
Of total splendor, the sudden exaltations,

Flash of a thrush, a rush of golden insight,
To be caught up in titanic light
As if one saw into the depths of things,
Yet averts the eye, to try yet further mysteries,

It is into this rich reservoir
Of knowing and unknowing I flash,
And shake high lightning spears of life
In the long combats of mortal strife,

Thrush song piercing human ills
With rigor and wrench so deep,
This glimpse is of an immaculate joy
Heart suffers for, and wishes to keep.

THE LAMENT OF A NEW ENGLAND MOTHER

Where have I lost my way among money and horses?
My mind is like the keen edge of a blade.
I am a Bourbon of Vermont. My children are prickly in the wheat.
In the castle of torment I swing in the winds of chance.
Do I dare my Adversary to duty past delirium?
Do I mock the green ancestors in my recklessness?
Is there any recourse when death has taken my beloved?
That cancerous fiend has made my assertions vain.
The world is rocking that most stable was.
My estates panic in the trembling of my will.
I am the crossed and vexed one, the soiled evidence
Of universal malevolence, guilty to have been born,
To marry, to bear children, to affront Providence with spleen.
I am in search of a soul revolutionized.
When shall I see the pure stars of my childhood,
When shall I trust in the love of my pure husband?
When shall I unseat my selfishness, my false debate,
And live in the rich simplicity of the earth?
The world has visited me with viciousness
And all my life is humorless and viscid.
I cannot cope. I am the lost cornucopia
Of June. Yet I seethe with rebellion still,
Daunting society in the mazes of my perfidy.
Let me go, Fate, and bring me back to douceur.
The graveyard on the hill that holds the bones of my husband
Affrights me with the rancor of life. My lovers
Have all gone into the garden. My richness of fantasy
Plagues society; I am a checkerwork of secrets
Knocking together in a burden of black action.
Wring me when my hands were dry, Life,
You wrung. Despair is noncapitulatory.
The delicate psychiatrist has become more delicate
For dealing with me. I improve his uncertainties
As he drives me further to the absurd.
Why should he be able to unlock my balance?

I am his high price. I had never dreamed, in the green
Wonderwork spaciousness and grace of my Springtime,
That it would come to this graceless nightmare and fact.
No restraint. I am the naked ecstasy
Of Fate. I thought I had reserves of strength.
I have nothing to live for but endless night
Without stars, moon, lovers, or a dream of peace.
So do I dream of a heavenly Adversary
Who will take me to peace, and he is death.
I call him then sweet names, and pour out all my love
And take in my arms this impossible savior
As a result of the cruelty of reality.
How should I do else? I am corrupt and unforgiven.
I have lost the long battle with myself.
My children suffer, but I do not suffer more,
Having suffered too much in the death of my husband.
Let the ordeal of life confine itself
To graceless fantasies; I incorporated reality
In the beliefs and charms of our society.
My studious Vermont of the sage ancestors!
My riches, my great inheritance, my satisfactions!
One by one, in deep and subtle wrenches,
I am reduced. All the glory is taken away.
My mind itself, now struggling in the balance,
Teeters debile, in idiotic frenzies.
I know that I am keen in shrillness and immediacy.
I am gone to carelessness, and survive by chance.
For what? My deepest accusations seem petty,
O that lost loveliness of the ethereal!
Should my mother accost me, I shall say
I am a mother. If my children, what
Can I say? To go with one's father is over,
My husband was shaken from life like a seed.
I stand alone, driven out of Eden,
Knowledge too much for humankind.
I in my borderline, adept and side-stepping,

Have gone over to the snaky wilderness.
I am the force that whirls me serpentine.
I look my fate directly in the face.
A revolver within my paleness and hauteur,
I commit myself into the hands of the State.

THE LOST

Most intimate, most far, most ethereally near,
It is you I write to, without a name,
Nameless and evocative and purposeful, blear,
Whose infinite life I esteem and claim:

It is you I hold most clear and most dear,
Most evocative and forever most pure and sure,
You who were destroyed by the rams of a tear,
For whom psychoanalysis had no prayer, no cure:

It is you who will read this in misery
I write for, you who may survive the volt fates,
And read the evolution of your stormy essence
As vessels diced and tolled in watery estates:

It is you who have escaped essential vainglory
And you who have trapped a final triumph of pride
At whose side in the purity of this clamant word
At the world's verge and judgment I rise, I ride:

It is you who drowning now were green with hope,
Whose life burst like a flare and dropped from sky,
Whose attitudes I espouse dark-heartedly:
It is for an impossible cause I would die.

Nameless my love, my loves, my many esteems.
It is all who are broken, who are nameless and reviled
I speak for in a language of the stalwart and kind,
In the redemption of forgiveness, as grace is to be mild.

You who were the image of the human fate,
You who shook, and you who arose and tried,
Who remember the savage depths of the world,
I speak for, reach. You shall not have died

Until each human heart lives for love alone
And every human spirit is enriched, occulted, blessed
With the hidden inherent spirit of the godhead,
Imperishable as the spirit of the possessed.

It is then, when love is victor and conqueror,
That, relaxing my revolution and astonishment,
Nameless my love, my loves, my many esteems,
I shall the gall, the spleen relax, take back, and repent.

MOMENT OF EQUILIBRIUM AMONG
THE ISLANDS

The sea repeats itself in light flourishes,
The southwest breeze-up of the midday
Is a lavish presentiment of possible danger,
Coves beckon as waves attack the prow
And slip past in stubby frenzies of loss.
Then we dare the open ocean; the green swells
We ride over with thorough, lordly motion,
Lovers of wind, sun, and the world-turn horizon,
And seek a new island, with a small spit of sand.
The anchor holds; we climb through contorted woods
Up boulders to an old granite quarry, whose
Dark, green, still, fresh water refutes the ocean.

It is the moment of looking down to still water
From massed granite blocks pleases the soul
With the hardness and fantasy of the world,
Before we must try again the gripping buoyancy
Of the salt sea, whose profound depths
Appear only to the imagination, while eyes
Survey the fresh roads the vessel walks
In triumph of buoyancy, delicacy, and strength,
As a philosopher continuing in the essential.
Then standing to the westward-closing sun
As the wind dies and waves grovel to stillness,
We reach at nightfall the landfall buoy of home.

AM I MY NEIGHBOR'S KEEPER?

The poetry of tragedy is never dead.
If it were not so I would not dream
On principles so deep they have no ending,
Nor on the ambiguity of what things ever seem.

The truth is hid and shaped in veils of error
Rich, unanswerable, the profound caught in plain air.
Centuries after tragedy sought out Socrates
Its inexplicable essence visits us in our lair,

Say here, on a remote New Hampshire farm.
The taciturn farmer disappeared in pre-dawn.
He had beaten his handyman, but no great harm.
Light spoke vengeance and bloodstains on the lawn.
His trussed corpse later under the dam
Gives to this day no answer, says I am.

CHRISTMAS TREE

Up a heavy wooded hill
A brother and a sister go
As on a new adventure,
Climbing through a foot of snow.

Their faces shine; their axe is gleaming.
All morning seems to be their nurture.
They inspect the winter world
As if they were out to conquer nature.

Hand and helve now have their will.
They cut a Christmas tree from earth,
Two children shouldering home the trophy
To give the tree symbolic birth.

Then worshipping, not knowing,
With lights, and gauds, and gifts, they play
Lightly, in their youthful growing,
Nor climb to confront divinity.

LOOKING AT THE STARS

Hung up there in the sky,
The old pendant stars
In metallic blue night,
Cold and clear, neat in November,

Chaste, incredible and inscrutable,
I am their daft onlooker,
To whom a half a century of life
Is as a moment of this night.

They opened to me when I was young
Infinite spaces to my longing,
Inevitable portents of feeling
And blessed my growing and my knowing.

Now I must make my own myth
Among the myriad hard stars!
Now I must see them as false lights
Dependent on my subjectivity!

When I am not here to see them
May young lovers come and find
In the rubescence of their flesh
The sable and cold limits of the mind.

And let them bend together narrowly
Under the vastness before the daylight,
Cling together in the mortal lot,
Witness the beyondness of their sight

And say, All you who lived before us,
You to whom the stars were impenetrable,
They are mysterious to us too,
We relive an ancient ceremony

Of our unknowing and your unknowing
And see them chaste, infallible, white
In beds of blue, as now our love
Gives prescience and influence to the night.

DREAM JOURNEY OF THE
HEAD AND HEART

My head, so rarely rent
For all the rending time does me
Into a dreaming vortex went
To see what I could see.

I wanted to go down those steeps
Into a place of the unknown,
For surely, I thought, my head
Would save me with strong bone.

I went downward, circling wide
In cold, diminishing cones
Until, when far away from warmth,
I walked on bare and icy stones.

They were harsh but jewel-like, thrice
I tried to turn about and go
But the fascination of profundity
Urged me onward, swart and slow.

The strictures came with further travail
As inward came the walls and ribs,
More brilliant seemed the lights
In tossing, skeletal cribs

Illuminating darkness with strange rites
As my balked walk was probed.
Endlessly I seemed to stop
And now I was disrobed.

Head, ravenous intelligence,
Help me out of this thin place,
I cried, but nothing in the head
Caused radiance in the face

Until the heart, with pity of nothingness,
Woke in a dream of grace,
When I rose back into the world
On sufferance of the human race.

WINTER KILL

Word traps catch big bears in silence.
They hunt the woods for years in freedom,
Keeping the counsels of the bees and snows.
Then, once unwary, a foot is caught in a trap.
The big black mountain comes atumble down.

His picture is put in the local paper.
The expressionless hunter stands in sullen pride;
A small son touches the nose of the brute.
The gun rests easy by the icy carcass;
People come to stare at the winter kill.

I would have him noble on the mountain side,
Roaming and treading, untrapped by man.
Man kills him only half for meaning,
Half out of thoughtlessness. The steaks
Are passed around as tokens to the neighbors.

Word traps catch big bears another way
When the meaning is total. The way a poem prinks
Into the heart from a forest hill
Is to have it in words, but never to have it.
Which is to say it is elusive still.

LATER OR SOONER

Whether sooner or later
Does not matter. I should late
Make some final statement
About man's fate.

But I am fiery-starred,
Burning in reality,
And therefore what I would say
Is not what I can be.

Too much, or too little,
Leaves something to be desired,
My ultimate fealty
To what life inspired.

May I bless the inexpressible
Beyond tooth and fang.
When man rose up, he loved,
And when he loved, he sang.

THE GESTURE

From the drama of horror and despair,
Out of the window, over the casket,
Young girls are bringing spring flowers
Carried and proffered in a spring basket.

So light a gesture in so grave a time.
I am one who, flailed and threshed,
Wishes in his power to understand
What it is that death refreshed.

An intimate gesture of young girls.
The flowers are laid before her white face.
In the mystery of their understanding
Reposes what we know of grace.

ULTIMATE SONG

It is too late for the ambiguous thrush
To sing in our garden, unseen though telling;
His wordless song, indescribably rich,
We cannot go on guessing in describing.

Nor flash his meaning back to China
Where love sat to a two-stringed instrument;
Nor bend to a blue tall night of Arabia
Where imagination feasts on the pomegranate;

Also, the Egyptian tombs are fragrant and dry
With aromatic spices of planned immortality;
The fallacy of duration is their stone spell,
Who felt eternity in sand, Nile, and sky.

Nor can the song of the mysterious thrush
Take us to England, whose elusive nightingale
Pre-empts many efforts to discover her there,
Though the searching heart think never to fail.

No, may the thrush among our high pine trees
Be ambiguous still, elusive in true song,
Never or seldom seen, and if never seen
May it to my imperious memory belong.

VISION

I

Two hummingbirds as evanescent as
Themselves
Startled me at my study window
As sea bells

Heard slipping through the fog,
Or yells
Of children down the block.

Phenomena,
Prolegomena,
They were

So sensitively sent
Beyond my pane,
Seen through it,

I thought the hummingbirds were angels
In a world of morning
And flowers
Soon invisible.

II

Again they come
Two hummingbirds,

But my eye,
Prejudiced to angelic vision,

Saw them not as brown,
Which they were, brown machines;

They riffle and rifle the flowers,
Sense-drenched in September;

Sense-drenched I exult
In their lithe bodices—

Where will they go, above
The frail duration of a flower,

Powerful in frailty,
Come October, come November?

III

The hummingbirds of hope!
Hope they are of all
That is exquisite and beautiful,

Small, round, and smooth,
Burke's three categories for the beautiful,
Held in an instant

Before the eye which triumphs
In realization and sings
Insistence all through the being

Of the rich delight of this seeing,
Sings of purity and power,
And hope that is highest

When visions of the earth
Like intuitions before birth
Sing, sing with the hummingbirds.

MAY EVENING

Long after our departure
Someone in a moment of significant rapture,
Seeing a boy beside a fountain,
Watched by an elder in a garden,
Will think that the past is the future
And the present is both.

We live in the imagination of the moment
When in a harmonious instant of apprehension
Subtle dreams are reality.
A boy playing by a fountain, unselfconscious,
A man watching him, studious in a garden,
Partake of immortality.

I am my father's father, or farther back,
Some enchanted man of the twelfth century;
I am Socrates' questioner in the agora,
I am a child dancing on the green seen by Blake,
I am all those to whom a moment has meant
A spell of rapture and a gift of grace.

The boy deploys from the playing water,
The man with his visions goes to get some coffee,
The incredible elan of the springtide evening
Lingers but departs; the graceful salute of the static
Moment of happiness and concord is given.
Fate outlasts the flash. Recognition was on us.

WAYS AND MEANS

I fight with the tools of the mind
But I love with the love of a lamb.
I believe in the objective world
To find out what I am.

I laugh at wolf-circling death
Because I know he is closing in;
No matter what stance I take
He will always win.

Will win! But I will abuse
His privilege with love of poetry;
Its power put on, and use
The subtle songs of immortality.

MEDITATION TWO

Style is the perfection of a point of view,
Nowise absolute, but held in a balance of opposites

So that for a moment the passage of time is stopped
And man is enhanced in a height of harmony.

He has purchased at a great price the gems of elan
In some avid precinct of his personality,

The price of years of doubt and belief, of suffering
The enigmas of the day, every hardy opposition

Of opinion, and every gain of hard-constructed good,
Music of furor, or insights passive and sovereign

When the clearest dreams are in a half-lit wakefulness,
When the best love is untutored, able to be blessed.

It is the style of the mariner proud on his vessel,
Who keeps a weather eye to the storm, but hopes,

Aware of the improbable, weather will not alter
From gentle zephyrs allowing him the spectacle of July

As if the afternoon were perfect and endless,
Porpoises in pairs follow the ship, and seals

Poke up their hopeful heads to see what trespasses.
The lobsterman is still at pulling his traps,

And far off the race jockeys on its summer errands,
Lightly touched with an ethereal evanescence,

Before returning to the home clubhouse and yacht club,
Inevitably pulling down the small sails at nightfall.

So should reality seem to be a style
Consummate and faultless, held in the hand

As the tooled wheel before the magnetic compass,
And all should be orderly in earth as in heaven.

But that we know the gale will rive us,
Years cut down our vanities, time unseam us,

Force throw the weak baby seal to death
On the rocks, the unexpected shock sink the vessel,

Or worse, to see the oncoming rollers and savage tempest
And know our doom forced against any wood or canvas,

Where is the style then for man the master of earth
And of waters, man who thinks to control his life

And to roam through the black new wastes of space
As if he had comfort in his small, cramping capsule?

Is there an outer misadventure or foul catastrophe
So malign as the malevolent sunderings of the soul?

For down in the depths of the heart's adventure
The evil in man since the loss of Paradise

And that knowledge which came in the Garden of Eden
When Eve offered man the fruit of the womb and of life

Has taken every stride with his heavenward hope
And locked his going in his ever knowing dualism,

So that from the opposites of good and evil, flesh and spirit,
Damnation and redemption, he is never absent

But truly is fixed in a vise of these opposites
Contending manly, forcing his sperm on children,

Unable not to start the chain of being again,
Crying out again and again when he sees suffering.

Is it not a provocation of the spirit of unity,
That, despite the ramifications of disparate phenomena,

Man seizes immortality on the instant
And can make his watery flesh seem permanent

In the magical power of a given poem,
In the working of paint, in the modelling of stone,

In the flash and controlled passion of music,
Is not the style of the man caught in his art,

And is art itself not a triumph of nature,
Before the worm takes over, before the breakneck tomb?

I sing the harmony of the instant of knowing
When all things dual become a unity,

The power of the mind to envisage singleness,
The purpose of the hand to shape lovingness.

If I sing Aeschylean right-mindedness,
It is of myself mostly that I sing,

Hoping the improbable advent of unity
Will triumph over the mocking dualisms

Which, each seeming real, yearly tore me
In the macerations of their blooded factions

As, whether to fly out, and shout with the government,
Or, silent as a crab, burrow in the sands of solitude;

Whether to embark upon the waves of chance,
Or reside in some closed nook of contemplation;

Whether to accept the brotherhood of the many,
Or live for the talents and the truths of the few.

So should style amplify and refine man's poise,
Be an instrument as lucid as the best of his knowing.

THE ILLUSION OF ETERNITY

Things of this world
In pure afternoons of gold,
And splendor of October,
Radiant air, still trees,
Give the illusion of eternity.

As if there were no suffering,
No ancient heart-ache of the being,
No tortures of the soul,
No struggle with mortality,
But changelessness, eternity.

A leaf falls here and there.
There are small birds a-chirp,
A chipmunk on a pine tree,
No cloud in the sky,
October afternoon, gold rarity.

Through the transparent air
Time is a kind of singing
In the inner being,
Acceptable singing,
Giving the illusion of eternity.

THE STANDARDS

I approach the standards of the air
With willingness to see them there.
Did I set them full and fair?
Can I change the standards of the air?

I saw the sparkles on the sea
As light toys, heavy imagery
Revealing natural constancy.
I did not know that they were me.

Great hieroglyphs of being
Ever appearing and fleeing
Are momentarily freeing
Truth's monumental seeing.

By the waters of life I stand,
Spirit in the sky, flesh on the land,
An absolute in either hand
No child can see or understand.

Time is the tone all fates foretell,
The birth cry and the funeral bell,
My will can wish, my psyche tell
That mystery surpasses well.

THE BIRTH OF THE SPIRIT

Some desperation of the sense
Has made me mad again.
In that madness I am well
And have the strength of ten.

My vision springs alive, replete
With life I am dreaming of
In spite of the nature of man,
A world of wholeness and of love.

I see a glory in the skies,
A brightness bearing down on time.
Out of my savage realizations
I live to make a saving rhyme.

I strike for order and for peace,
I state for grace and harmony,
All natural beauties of the world
Of bird, of beast, of sky. and tree.

When this madness comes to me
I have the strength of ten,
My spirit comes up clear
As that of other men.

When I land down among events
I cannot see with truthful sight
But when I am flying high
I see the orient light.

Blessed be madness and power
For they so force the earth
That spirit is a red flower,
The pure red flower of birth.

EXTREMITY

The spirit cried Go up higher
The Devil said Jump in my fire.
In the chaos and strife
I want one poem as hard as life.

I want the firm middle ground
That in truth I never found
Lest another extremity
Catch me and be me.

'MY BRAINS ARE SLIPPING IN THE FIELDS OF EROS'

My brains are slipping in the fields of Eros
My contamination of the absolute
Intrigues me with the confrontation of imperfection
In which I veer from the cradle to the grave.

I would be a blade of grass stabbed by sunlight,
I would be a bell tolling soft evensong,
I would be as a lover come to the imagination of heaven,
I would be the pure ego in its evanescence,

And let science roar in the headlines,
Let spacemen eventuate in distances,
Let the wonder of the passionate truth-finders
Extol the ecstasies of the farthest choice and reach.

I stand in my times as a snowflake
Remarkable in its ability to perish,
And as a song of the woodland thrush among pine trees,
Exquisitely attuned to possibilities of eternity,

Which is to say that, strong as mountain
I brood over the landscape without remorse
And announce the heart of man like an avalanche
Exciting my terror and solicitude
And the dreamy villages of future time
With joy and love, implacable and certain

So that the dreaming futures of mankind
May be replete with joy and love, implacable and kind.

REFRAINS

The temple of the body crumbles,
The vision of the mind remains.
In the midheat of man's troubles
Poetry plays its fond refrains.

There is soul for song at noon
Before the darkening of the larches.
Shake then shake historic doom,
Peer through interstices of the arches

To some gloom so deep so pure
Your being cannot pierce its being,
You are a stranger where you are,
Nor could you see where you were seeing.

So go the years. So stand the times
And body falls and mind remains.
Fathers seek their templed sons
And heart and head hear old refrains.

TO HARRIET MONROE

Then I came to Harriet Monroe,
I was so young, I was so light,
My spirit leaped so far,

I walked with her down Michigan Boulevard,
That beautiful head, that delicate balance,
The look of intelligence, she was so short,
O how I admired her worldliness!

Later she rubbed my words, her audacity
Assaulted me in my citadel,
She dared to change a word of mine,
I fought her, we broke.

Later we made up and she died.

'WHENEVER I SEE BEAUTY I SEE DEATH'

Whenever I see beauty I see death
Because she died when she was beautiful.
This death is everywhere, and so this beauty
Pervades the universal atmosphere of time.

I am possessed by ancient harmonies,
Am moved upon by joy and pain together,
And I in turn move to a ritual of knowing,
You see these words as inflown from the stars.

Now the birds cry, and now the children call,
While all about in passionate acclaim
Truth pierces matter where the heart is rent,
Mind pierces spirit when all loves are lit.

RECOGNITION

I admire the roughness and brutality of the sea.
It cuffed a corpse around like chaff.
There are the sophisticated, those who laugh.
But the brutality of the sea, O dear to me.

I love the grossness of the power of fate.
If we have assassinated a President,
Centuries ago flowers to such were sent.
Grossness, itself, has never lain in state.

When belief in comedy has put me at a loss,
And a rich thrall has made me wish to kiss,
I think of the purity of a child, and in bliss
Make love to Christ, cinctured, on the Cross.

Fate has reached me through iron mail.
Brutality, grossness, and lust to dust
Must come, as upon the just
And the unjust fall rain and hail.

But thou my sovereign of the pierced side,
O thou deep knower of the true,
It is from your side my poems grew,
You who rushed ahead to Eastertide!

OPULENCE

Nothing is so magnificent
As the sun descending,
Copernicus-gold over the horizon,
With birds singing in the pine trees
When it is rich summer, when June
Has on her iris finery
And peony-bright, hesitates good-bye.

Nothing is so magnificent
As the full mind, stored with summers,
With age approaching,
The sun standing over the horizon,
Wonders yet unknown, love not refusing,
The world all a visionary
Guess, unspent clarity.

MEMORY

There must be time when you, too, dream
Of the perfection of the evening, music
Glancing from the resistance of the curtains
To seal the heart in certain silences,

When the possibilities of an earthly perfection
Play among the shadows of the curtains
And, invisibly, descend upon the flesh
Suspense of being, benediction and essence.

THE VASTNESS AND INDIFFERENCE OF THE WORLD

I saved a chipmunk from a cat.
But does it matter?
The cat was too dandiacal
And kept the struck chipmunk
Waiting in its mouth.
I thought it maniacal,
As well as in patience angelical.

Meanwhile, a wife was talking to a daughter
In serious tones of manners and what was the matter
With the world, and to young tears laughter
Dafter would seem than any animal slaughter;
Indeed I was held spellbound
(Somewhat hellbound so I felt)
Between the drama of the larger and the smaller animals.

Meanwhile the cat was scared off
And the bitten, smitten chipmunk hobbled,
Without thanking me, quickly to some under bushes
As if he did not know how close to death he was,
Who had been held patiently in the cat's jaws
While she decided how long to wait, in her
Elegance, to dandle him another blow toward death.

I meted out my justice, I became the justicer
Of nature, I became a conscientious objector
To cats, and I obstructed nature. Obtrusive
I. I by instinct defended the weaker animal
And saved a chipmunk from a cat.
The human dialogue went on, however,
And the evening cocktail party, pressed with

Significance, and oppressed with ignorance
Of what I considered real, not seeing it,
Could not have cared less
And was not all that impressed
With the indifference of the cat to kill at once
And the slowness of the chipmunk to revive itself
But for the poet's intervention, which is to save the world.

HILL DREAM OF YOUTH,
THIRTY YEARS LATER

The river of sweetness that runs through the meadow of lies
Meanders through the providence of summertime.
Seen from a hill, where in my youth I contemplate
The vastness and resource of the imagination,

I think to dream the consequence of being alive.
In the command of my faculties I observe.
I see the richness and grandeur of the summertime,
The river of sweetness that runs through the meadow of lies.

There is a reality of wealth, the wealth of nature
Appears to my sight as beneficent to my fingertips.
I finger a flower and hold the whole of nature,
The river of sweetness that runs through the meadow of lies.

Live dreamer, living in dream, I note from the height
A summerhouse in the meadow by the running river.
Are there people there who mate with my beautiful desire?
Can beauty be devoted to the purpose of life's enchantment?

I see in the summerhouse the quarreling couples,
The desperate chances, the hardship and the ill fate
Of a dark nature hidden by the river of sweetness,
Through open, cut windows I hear the stomp of tragedy.

The dialogue is prompted from the wings.
We have seen our friends destroyed and ancient Satan
Comedian to our enterprise. We take our cue
From circumstance, while time hardens the soul.

Yet, I would keep life at a certain distance,
Out of respect to it in the depths where it lives
In the heart beyond belief in appearances
Of the river of sweetness that runs through the meadow of lies.

I dream Plato's dream of being from a high hill.
I am myself entirely happy in the summer intuition.
Young, I contemplate the territories of the magical,
The river of sweetness that runs through the meadow of lies.

WHY?

A poem so hard it has no heart.
A poem that shatters diamonds to dust.
No voice from the clenched throat,
No flickering light of hope.

The poem of the young lovers
In the comedy of their exercise,
Seen through the lens of age
In its comic misadventure.

Why do not skeletons arise
From the muck of our philosophies
And tell us what to believe
In the thrust of April skies?

R.G.E.

A new desire to understand
Took hold on me and drove my hand
To prove my own identity
By digging letters on a tree.
'The maple tree will bear my scar
When I have grown and traveled far.'
I said it blithely as a boy.
I carved the wood with eager joy,
Gouging my three initials deep
And deeper still, so they would keep.
It was an impulse when alone
To clarify the vague, half-known

Surmise I had that I was part
Of trees and knew them in my heart.
It was believing that the tree
Would be forever nearer me
That made me cut away the bark
And trench the maple with my mark.
Brutal the one who feels no sting
In injuring a perfect thing.
The maple felt a primal shame
To bear the imprint of my name.
It only wanted sustenance.
It felt no need of permanence,
Because it knew the earth and stood
All day in silent brotherhood.
But I was racked and torn. Hot pain
Shot through and tingled in my brain
And would not cease, because I knew
Truth that hurt me through and through.
Never can my human will
Be calm and tree-like, free and still.

1918

TO THE FIELD MICE

Come small creatures of low estate, friskily moving,
Make sally along the stone escarpment, toward evening.
The stones do not move in their millennial inability,
But you move, bright-eyed creatures of the height of summer,
Seem to scamper on errands of urgency,

You show yourselves in flashes and dart back in the rocks,
You look out with alert eyes, small in the large, still rocks,
Instinct with life in the short time of your devotion,
I behold you and love you as opposed to the rocks, dashers
For food or whatever you are doing unknown to me observing,
I am god-like to you, you do not know that I exist,

See I sit still, I watch, I observe your cautious walkings
As the late afternoon changes to early evening,
I wish you well whatever may be your purpose.
Above you stand the flowers, above them stands the sky.
It is the height of summer, I watch the grand occasion
And give you, white-footed field mice, my fidelity.

THE ASSASSIN

I am that vile worm of Satan
Sent to kill the beautiful.
I have destroyed everything that is good.
I have drawn over my head death's hood.

In the dark of my brain
Was a primeval stain.
I destroyed what is sacred
Out of primitive hatred.

Grace fell from my face.
Of love I had not a trace.
I was able to kill
By my malevolent will.

Thoughtless, abstract, evil,
I was agent of the devil
Who overwhelmed my humanity
And totally blinded me.

Now whatever gods there are
See in history from afar,
The repetition of the stroke of Cain
Come to the world again.

BALL GAME

Caught off first, he leaped to run to second, but
Then struggled back to first.
He left first because of a natural desire
To leap, to get on with the game.
When you jerk to run to second
You do not necessarily think of a home run.
You want to go on. You want to get to the next stage,
The entire soul is bent on second base.
The fact is that the mind flashes
Faster in action than the muscles can move.
Dramatic! Off first, taut, heading for second,
In a split second, total realization,
Heading for first. Head first! Legs follow fast.
You struggle back to first with victor effort
As, even, after a life of effort and chill,
One flashes back to the safety of childhood,
To that strange place where one had first begun.

THE ENIGMA

Mine will last though others fall to dust.
He began to row across the flaming lake.
Acrid fumes were sweet as the unexplained.
Tears ran down his face all day
As on the whirling came, as come they must.
In his gutted heart bell's memories would break.

Mine will stay though others live in hate.
The lake was vile with forms tormenting mind,
The fumes through tears were dear reality.
They came believing but they faded in the maze
As he rode on time's mystery of fate.
Change as they would, there was no other kind.

Mine will survive though others fall to dust.
He neared the dark escarpment and white pall,
Crying in a wilderness of forms,
His head fiery, his heart hearing ancient bells.
They came to judgement, as come all must,
He thought he felt the meaning of the fall.

It had to do with struggles in the night,
With breakless Adam and still breakless Eve,
The ever oncome, the will to row in fire,
Bells bringing memories of helps long gone.
Though he hailed them, hurled them out of sight,
His would survive, though never to him cleave.

THE HAYSTACK

In memory I see a youth
And a girl looking into each other's eyes,
In the surge and height of their life principle,
Which they do not understand.

On a northern lake in a rowboat
They catch a glittering fish, slight
Catch, they think nothing of it.
Grace is theirs, youth's insolence.

Back at the house an old man
Drools and cannot speak a word.
They look at each other. They drool.
They fall together behind the gray-green haystack.

SANTA CLAUS IN OAXACA

Nothing seemed so incongruous
In this Christian country of Indians
In bright clothes, Indians part Spanish,
And tourists neither Indian nor Spanish,
In the warm dusk in a place of bells
When the cathedral rips again the harsh sound
Of every quarter hour, and then the full hour,
Next to the Marques del Valle Hotel,
And in the square the noisy band goes off
Like a jubilant series of firecrackers,
The firecrackers shot off by Mexicans young and old,
Where every breath taken is compassionate,

Nothing seemed so incongruous
As to see Santa Claus in the hot lands
In red cotton garments, trimmed in white,
His bearded face impersonal but appealing,
Walk awkwardly through the square of Oaxaca
Followed by popping strings of boys and girls,
Mothers with babes mangered in red rebozos.
Where are you going, Santa Claus, walking?
Are you going to the ruins of ancient Mitla?
A gentle Zapotec explains the tombs of Mitla.
This Zapotec survives, but gone is the last fierce Aztec.
Are you hastening to see where the future would go?

LOOKING HEAD ON

I sit looking at my features.
They are hard as rock, they are, in fact,
Fired in clay by Winslow Eaves.
I look at the features.

No man has seen his own face,
Except by mirror or photography.
We, who live so much within, look out
From ourselves, through the fires,

But cannot see ourselves as others see us.
I look out. I have been looking in
Through decades of the world's history.
How dramatic to look at your own face

Hard as rock, fired in clay, as the sculptor
Sees it. Which, then, is reality?
Is it the flesh that pulses, fails?
Is it the rock-like likeness, clay-lasting?

Bold is the opposition
Hard to define in a day.
We leap in virility, but
There is the slower finality of clay.

SOLACE

It is certainly the mystical hour of the day
Before the sun descends, with
Birdsong among pine trees,
It is certainly this hour above others
Mutes the action, the passions of the day,
Holds in suspension the terrors of life.

This June hour of splendor and quietude,
Before the inevitable darkness
Covers all in sleep and nothingness,

It is this hour of enchantment
I celebrate, late sun through pine trees,
The river rippling quietly below,

Solitude in the atmosphere,
Quiet, immemorial calm,
The blessing that is not a total blessing,
Solace to the enigmatic sufferings of man.

EVIL

When I entertained evil
I played upon him as if he were good.
At my red banquet table
I set before him peppery food.

I thought he was king of the world.
His elegance, his subtlety
Were without question in my mind.
His sensibility was exquisite.

He seemed like a devil incarnate
But so much like my friends, myself,
That I recognized a hidden truth.
At a banquet table nothing offends.
We talked of the affairs of state.
Should one turn the other cheek?
The idea was that to lose face
Was a shame in being weak.

None thought to kill was bad.
The pictures of the lacerated Vietnamese
Were somehow not to be mentioned.
None thought that any here was obese.

There is a certain delicacy
In what to say at a dinner party.
The idea is to accept man as he is
And rejoice at eating hearty.

Now I opened a bottle of Rosé.
It was a symbol of relaxation.
We were all feeling well
And I offered a toast to the nation.

I said, let us drink to freedom.
This seemed brightly reasonable
As everyone around the table arose
In our state of hedonism.

We drank to the glory of our state,
None thinking this uncouth.
We drank to individual aims,
And to the complexity of truth.

Our guest, a ruler of the world,
Was delightful, polite. I saw his bloat.
He said that evil was the greatest good,
My imaginary bullet through his throat.

MARRAKECH

The dance begins with the sun descending
Beyond the Koutubia in Marrakech so ancient
And so fable present: the old tell fables
As the Moroccans listen in eyed attention
In the marketplace of vitality and veiled women.
But the dancers, O the dancers, priests of devotion
From the high Atlas mountains, perhaps twelve,
A boy of ten, shining men under thirty,

Their feet bare and hard on the bare, hard earth,
Begin to dance to two taut goatskin drums,
Beaten with hard crooked sticks, thin in the diameter,
The one fiercely antiphonal to the other,
Together beating compelling rhythm to action,
Action in the flow of the loose, ancient garments
Of the men as they strike in both hands
Double metal castanets in the dry, high air of late day.
One would come forth, loosening his devotion,
Gyrating and flashing in compelling immediacy,
Total in ecstasy, overwhelming the senses,
And fall back, and another would step forth lightly,
Deftly begin his interpretive energy, total
Devotion to sound, rhythm, style of the dance,
And fall back, and his fellow come out before
To outdo the predecessor, turn and leap and gyrate
In ochre ambience, and the drums' insistence
Proclaiming efficient animal action,
The passion of primitive man exultant,
And each came forward, each leaped taller,
Flashed lower, turned subtler, rose higher

Prolonging intensity of animal strategy,
A rapture of magnitude pervaded the air,
Sound and sense reached to magical ability,
One drummer on his knees, the drum head vertical,
Beat out the passion of ancient centuries,
Appeared the thronging nature of tribal power,
And the laughing gods fell to the earth, spent,
And the high heart bent down with them to the earth,
And the heart was raised up to the Atlas mountains,
In the superabundant, delirious air of sundown,
And the laughing gods fell to the earth, spent,
And the heart bent down with them to the earth,
Cleansed in the nature of rhythm and rite,
The dance was a thing in itself triumphant,

Music and dance the perfection of the free,
And before one could think of the meaning
The driven passion of the drums begins again,
O ancient Africa, O tribal ecstasy!
A dance of six hundred years on the same spot,
They come out again instantaneous and eternal,
And leap and turn, passionately leap and fall,
And all are made whole again under the red sky
And all is made whole in the heart and time.

LIONS COPULATING

We caught lions copulating on the plains
Of Africa. Our landrover closed in,.
I pressed so hard on my 8 mm. color film
I almost lost this gigantic naturalism,

Trying to preserve it for my friends and astoundees.
But saw the King of Beasts with his head high,
His mane imperial, no expression on his face,
Prodding in and out of the great female

As if he were a schizophrenic dualist
And had to put up with his baser nature,
For his great face had no expression at all
While his lower being worked mechanical,

Then he fell away, and stood off, and lay
His full length on the ancient earth
While the lioness with a sumptuous gesture
Rolled over as I have seen other females do

In the perfect surfeit of her animal nature,
And took ease as if nobody were looking on,
And after an interval of valuable rest
These great beasts of the African wilds

Stood in their historic posture of superiority
And ambled across the limitless plains in silence
Without a thought of the lucubration of man,
Trying to signify their big natures in empathy.

Nairobi

THE IDES OF MARCH

As I was riding through New England
Along about the Ides of March
I passed by an ancient graveyard
And looked hard at the stones:

It was the deepest look I ever gave,
It was a look I had given long ago,
The most intimate and the most personal,
For I wondered what I signified.

I could not see my next connection,
Nor any of the loves I have in life
As divorced from that immaculate glimpse
Of the permanent end of human being.

The car bore on through the countryside,
Plunging in meshes of society,
We were going to meet our friends in the world,
In the next connection of belief and delight.

We were going to meet the world incarnate
Without reference to the rigors or Ides of March;
As we slid through the freshening Springtide
We faintly apprehended the fatal glance.

It is that glance of the eternal judgment
Of the silence of the manifold gravestones
Frightens me as we seize in love and belief
The love not to ask what is our ultimate end.

As I was riding through New England
Along about the Ides of March
I was the essence of the questioner,
Only the glance of the gravestones answered me.

A WEDDING ON CAPE ROSIER

Today there is another marriage
Of the very young by the side of the sea.
The caterer is so pleasant he seems one of the guests.
The parents are divorced but the yachting young
Take communion along with their vows
In a forest chapel looking out over the ocean.
A small boy strews rose petals one by one,
So delicate and beautiful an operation
As to defy the desires of poetry,
While a dog comes into the service
With rude life to remind us
Of the charm and stability of the animal kingdom,
Which we deny in singing
The praises of life everlasting,
That deepest wisdom of centuries of Christendom
Which speaks for the vanishing of everything present
Although the young couple in their youth's beauty
Could not be more evident, capable, or potential.

Father Emerson puts out the candles.
His father knew Emily Dickinson.

It takes so short a time to get married,
We noticed no change in the tide.

ON RETURNING TO A LAKE IN SPRING

When the new frogs in their exuberant arrivals
By hundreds raised their voices in lusty croaks
As I walked up to my knees in reeds among them,
Wading through the wet strong forces of present nature
As if I felt for the first time a divination
So powerful as to shake my frame beyond words,

And only the small frogs could speak harsh articulation
Of the pure force racking them to ecstasy,
When I strode like a god among the small,
The bright movers, the true, walking in triumph
As a king of the frogs, glad to be among them,
As they touched my legs as I moved along,

I remarked their destiny, their birth and death
As one sure of mine, the minimal existence, here in
The great opening of the world again, sure as I was
Of the summer life, in full sunlight, of frogs and men,
I was equally sure of the fall of the year, winter,
The razoring sleet, and locking ice and snow.

I returned to the picnic on the hill exulting.
In our party a young woman moving in her youth
Seemed to jump at the glory of the springtime;
Nevertheless she did not speak of the spring peepers.
A month later, in a southern bog, she slit her throat.

THE EXPLORER ON MAIN STREET

You will see this big-jawed man walking up and down
On Main Street in a small New England town.

He was a Franklin for exploration in the arctic,
He was a titan among the Eskimos.
In the long walk from the iced-in ship,
In the death-dealing hostility of nature,
He learned a particular, savage subtlety,
Do not eat the flesh of a lean man
For you will die of protein poisoning,
But, when they fall, chew only on fat flesh,
The fat and the lean make a wholesome diet.
You may not be able to walk another day
In the ice-antagonism, the bitter wind, the wail
Of the wind's savagery in the long, lone waste lands,
But that you feast on your fellow, fat of the land.

The big-jawed man was walking up and down
On Main Street in a small New England town.

At the dinner table he was absolute dictator,
Old, alien, able in his truth,
Demanding to speak his intractable reality.
He held companions speechless with his speech,
A long gauntlet of authority,
Side-stepping the mind's ultimate crevasses.
We respected the grandeur of his age,
He was a hero of the northern wilds,
His was the human spirit at his highest peak,
He ate only the flesh of animals, no vegetables.
Yet when I said, George Bernard Shaw
Lived ninety years eating only vegetables,
And no flesh of animals, he would not comprehend
This, saying, eat only the flesh of animals,
The fat and the lean, and eat no vegetables.

I thought he had more to explore on Main Street,
In a small town in New England,
Of the mysteries of nature, the subtleties of psyche,
Than he had in the arctic wilderness
When his great strength seemed essentially simple
As he manned his way through devastation to survival.

I thought how strange it was to be world famous
Only to show man in his strong, basic survival,
When the struggles of the soul, imagination's roil,
May yield some port beyond the moon and stars.

SANDERS THEATER
The Phi Beta Kappa Poem, Harvard University, 1967

Into this narrow vault I came when young,
Rude with enthusiasm, my need girt,
Due to deprivations of the soul,
To discover the grand scale of life.
    In the corridor as I came through the door
    I thought I glimpsed the clarity of God.

It was young dreams and God too bright in the glass.
Surely within us we carry what we would see.
My trial was then as now to compensate
Inner belief against exterior reality.
    I moved to a pew to hear my superiors
    And sat aglow and lithe among my peers.

I heard the sermons with the serpents' tongues.
The vast confusions of the world were here.
But most I saw man's enemy is man,
And then I saw my enemy was time.
    In a kind of blindness never understood
    I made and sang my freedom-breeding song.

Fear shook my bones as if my skeleton
Were played with by some mighty force at large
I could not cope with. Gladness made me run
Hard on the world with undeviate will.
　　It was my psyche led me through the world
　　As if hope were not hopelessness to feel.

As joy touched my cheek and made me lightsome
Death swiftly beckoned to the port of peril,
Would I have time to make some resolution
Of the violence of man, the purity of the soul?
　　War finds our miserable secret out again
　　As we kill each other with negative capability.

And by what clarity, and by what poise
Shall man see again his radiancy,
Has he not died enough to be reborn
By love to a life beyond savagery?

Shall man destroy the face of the earth,
By universal hatred bomb himself to death,
Shall he still lust, erect evil
And shall madness overwhelm our reason?

I see space travel as instinctive will
To solve our unredeemed problems in the heavens
As if man were reborn again as a child
To the limitless delights of the imagination.
　　In the old poem the cow jumped over the moon,
　　We plan with luck to hit it on the nose,

And go off on further sidereal exploits
To escape the confines of the population explosion,
For surely if we cannot stand up on the earth
Some mutation will provide heavenly exodus
　　And we shall be as we have never been,
　　Things will not be, as always, what they seem.

297

But new. How shall I make a mythology
Of our gross and stubborn naturalism?
What words are there for our new energies,
And how shall we speak to the generations of the young?
    Shall we say science is a fairest flower
    In the gardens of our inconstestable ecstasies?

How shall we use a computer mind, how ever,
To walk in the comprehension of the intellect?
Do I have to be ravished by the psychedelic?
Do I have to pray that I shall be undone?

When to the powers of song spirit speaks
There is a source of spiritual unison
Which links the songs of the centuries
To the anguish of the individual heart.
    The tides coming in and going out
    Are mysteries to the subtle understanding.

A ritual exists in which existence
Measures itself against its limitations.
Suffering and indecision are requisites.
The cliché of the heart knows no deceiving.
    The richness of life is beyond saying
    And the loss of time is beyond redeeming.

And the animal energies of our delight
Pass over us as years pass and are renewed.
Another summer comes. But we have seen
The last summer of the life of friends.
    We have confronted the first mystery
    Of our own light gone to brute darkness.

I prostrate myself before the old mythologies,
Believe not one of them, and demand a myth
Of the original spirit of the American ego,

A world-poetry of the international psyche.
    I call for the love of man in every man,
    Brotherhood from Iceland to Dar-es-Salaam.

I call for passionate love and passionate care,
For lack of violence and for love of peace,
For love of poetry as inner being,
And for freedom as inviolable need.
    In poetry of the innermost heart
    Of instinctual life shall be the spirit of freedom.

I shall believe that man will exceed himself.
What future will arise centuries hence?
Shall man stand, as in the tarpits of Los Angeles,
As a lost animal among the galaxies?

To this narrow place I am come again
Through millions of deaths and millions of rebirths
Teeming with desire still to penetrate that light
I saw alive on all things in my beginning.
    It was half a life-time ago I stepped
    With light foot in this adventurous hall.

And now I know that imagination
Is freedom and that poetry is praise,
I cope with man and am consumed by time,
God stains the glass at the back of the hall.
    I walk out of this hall passionate as when young.
    I came into it, I go from it in mystery.
    Coming or going no better word than mystery.

THE YOUNG AND THE OLD
(*For W. B. Yeats*)

Will we make it to 2000?
1984 slides along ominously,
Now murder and assassination assail free Canada.
The center is not holding, Willie.
I sat beside you when the course was young.
I knew, old man, you would tell the truth.
How the young laugh at your singing Byzantium bird.
Ventriloquists all, no golden bough for them.
Who would want the intellect, that great stride,
Only to reach a tinsel bough. The drab
In the ditch is better, you thought of that,
The *Purgatory* knife the sharpest you ever honed.

We are easy riders to the fields of grace,
A bombshell in the gut.

No reality but in the spirit.

OLD QUESTION

I should walk maniacal
In the diabolical
But that reason should
Hold me to the good.

Why down in hell
Do men feel well
And why in madness
Know gladness?

We dare to tear
Ourselves apart
To know the true, the rare,
For the sake of art.

We are so in love
With life, that only death
Is strife
To be worthy of.

JOHN LEDYARD

Only death remains
To tell us
How great we were

Great with life,
The stone entablature
By the river

Tells us
Of a youth who made a canoe
From a cedar,

Descended the Connecticut,
Discovered far places,
Died the great voyager

Far from home,
Lost in Africa,
Youthful, valiant, destroyed.

That was in another century.
The incised bronze fades,
His name only remembered

By insignificant lovers in the Spring
Who read his story
And clasp each other,

Amazed at intrepidity
For all they want is themselves,
His valiancy hallucinatory,

His trial of the world incredible,
Their heroes themselves,
They only want to clasp each other.

Only death remains
To tell us
How great we were

Speaks the voice of the voyager
From fading bronze letters,
Great with desire.

VAN BLACK, AN OLD FARMER IN HIS DELL
Times of Archaic Splendour that you saw

There are times of archaic splendour that you saw
That when you saw them could not show
The unexaggeratable meaning of their drift
Which appeared in memory with unalterable flow.

You saw truth, you saw reality, the whole scene
Of man on the face of the earth, holding a rake,
And with muscles in his arms as big as baseballs
He pulled the gauze of summer to thick windrows.

It is his care that matters to us in the end,
His bent stance in the glaze of ancient sunlight,
The grueling work as if he had the mastery,
It is his lean tough luckless finicky endeavour

In the swift downfall of the year in August
Makes pathos grand, grand against granite cliffs
That tower above him inhuman, big as time,
Methodically he pulls on the rake's loose teeth,

Up on the cliff at the stealthy end of day
A buck leaps from rocks, held in sky light
An instant, so magical, so graceful, so final
He knows he has seen the glory of the world,

Will never be able to put in words this vision,
A throw of the eye beyond the field of raked nature,
A dazzle of the sense, archaic splendour of action
Lost forever: the buck leaping, the man gone underground.

FROTH

When the sea pours out under the bridge
Over the dam at every outgoing of the tide
There is ravishing serendipity of froth.
I stand on a hundred-year-old millstone.

Froth gathers on the turbulence of the waters
Abundant, fresh, puffed up. For hundreds of feet
Small cliffs, bluffs, towers form and seethe
From the turbulent center outward gliding,

Improbable as heaven, like a child's view of heaven.
How could anything so frail seem so permanent,
How could anything so sheer evade destruction,
These are things of imagination floating slowly away,

A white suds against blue water, white over blue,
A too much of nothing strangely become everything,
Froth dominating the presence of the world,
Cliffs, bluffs, towers and castles passing by.

It is good that everything has turned to froth,
Magical froth of truths going out to sea,
While I stand on the millstone solid and gray
I hear the airy shapes whispering infinity.

THE SWALLOWS RETURN

For five years the swallows did not build
In the treehouse near the door facing the sea.
I felt their absence as furtive and wordless.
They were put out of mind because they had to be.

Then they came again, two males attending one female,
Skimming in the late afternoon gracefully, ardent
And free in quick glides and arcs, catching flies on the wing,
Feeding their young in the house safely pent.

It was mid-summer, the time of high July,
Their return as mysterious as their former leaving.
They presented the spectacle of orderly nature,
Their lives to some deep purpose cleaving?

At night there was clamor. When morning came
The ground under the house was littered with feathers.
None knows who was the predator, but death
Is available to birds as to man in all weathers.

THE WEDDING

With a southwest wind blowing late August,
Sally held a young seagull in her two hands.
With a love as supple as youth itself
In an ascending gesture she released the wild bird
In a moment of consummate grace and communion.

For an instant the girl and the bird were one.
Then the wild bird went to the wilderness of air
And the young girl returned to the confines of mankind.
Her gesture was like a release of the spirit
Held within the bounds of the corporeal.

Now we assemble on her wedding day
With no direct analogy in mind.
Bride and bridegroom here throw off the past,
But do they? Does a spirit of freedom hover
Over the future from the hand-held past?

TO KENYA TRIBESMEN, THE TURKANA

I love the Turkana tribesman who gave me his cane
Made into an eyelet of parti-colored wood, at Turkwell
On the Lodwar, southwest of Lake Rudolf, near Ethiopia.
I subscribe to the thin lips, the grace of the thin man.

I am beguiled by the stately grace of the young, barefoot girls,
Their naked feet held to the earth with moving cunning,
Each one Helen, immortal Helen, sweet Helens of Africa,
Tall, stately, naked, they were innocent of their glory.

The naked men with their spears seven feet tall,
The energy of tribesmen under the ancient trees,
Feathers arising from their heads, tattoos on their backs,
None old living, shining with tribal vitality.

What is this, old Death, how they have betrayed you,
I thought, incandescent creatures of an hour,
Here is no death, here is primitive lust, lust
For life, lust alive for thousands of years,

I am witnessing immortality,
They have no sense of a race with death,
Which is the surfeit of my Western poem,
They take me back to some immortal joy.

Boehme, Blake, Beddoes, be with me now,
You strove to know the truth of the real,
You would have loved to see the sights I saw,
Loved to know the feelings that I feel.

KINAESTHESIA

Sometimes I play the music of Charlie Byrd,
The dark rush of horses over the plain,
Confuse him with Charles Ives, fret of a flute,
The strings, the breath, and the interlocking time
Befit an inexpressible richness, lost unless
I try with poetry's imaginary gainsaying
To weave imaginable, placating harmonies,
Saving the mind for its penchant for newness,
While in the background are African Turkana shouts
Of the dance I saw under the thorn trees, ecstasy,
The brute mind of Beethoven breaking bonds
To say in the last quartets what I feel now,
An inexpressible richness, the austere, the sensuous conjoined.
Now the full fingers of Charlie Byrd play over the strings,
The austere face of Charles Ives strives from the record
Jacket, night flows under the heart in rivers of dream,
I am thrown along the decibels of the ages,
Music is the keys to the stars and the sea,
To the individual hour of the heart none shall forget,
Byrd and Ives joined, improbable, exquisite.

THE ANXIETY I FELT IN GUANAJUATO

The anxiety I felt in Guanajuato
In the second storey suite looking out
On the square to silhouetted theatre statues

Was a shaking and a reeling,
I saw a small boy leading a blind man
From the square down some ruinous street.

It was the blind man pulled by fate
My anxiety reached for. It acted
Like an incurable dictate.

The statues of the great actors,
Greater than life, were sublime
In their plaster indifference.

The small boy pulling the blind man
Put me into a frenzy of belief.
I watched, and did not move.

Deep down in the hilled cemetery
There are the corpses of Guanajuato,
Each held in his immortal gesture.

TRACK

Forced as I was, I ran a race with death.
I could not run a line as fine as his.
I ran it square, I ran it straight, but death
Was always out ahead of me, the winner.

As in a relay race, I strove to better time,
Hand on the baton with enthusiasm, spent
At the last instant, but adversaries ahead,
There were always adversaries ahead, cutting time thinner.

Now I walk as nature tells me to walk.
I like to think there is no competition.
I am myself. Whatever I am I am,
An end in itself. But death wants a new beginner.

He begins to loiter when I think. He slows
To notice and to savor my philosophy.
He cares for me too much, I think. God knows
What game he plays with me, Paraclete underpinner.

THE BOWER

Under dense fir boughs
On a carpet of pine needles
In golden afternoon
He came as to a cathedral
For the laying on of hands.

She came like the heart of summer
Beyond the reach of words
To exist in naked splendour
Where pain is forgotten,
To silent pleasure.

These two, realists of life,
Without a word to say,
Entered into the communion
Of old centuries, far places,
Zephyrs pleasing the sanctum.

It was a redemption as ancient
As lovers ever tried to imagine,
It was what battlefields are for,
To disclaim the evidence
Of the shooter and the shot.

They never thought of the suffering
Of the gross battlefield,
Under the serene pines
On a bed of pine needles
They never thought of wounds.

They never thought of useless death
Of boys killed in Vietnam,
of the bloody passion
Of man to maim, to make blind,
The blood-drench of mankind.

They were in Eden, two
Enacting the high dream
Of purity beyond the evidence,
Their flesh effulgent
As in Rubens, Velasquez,

Zephyrs playing among the boughs
Nimble in the lightsome air,
Jets of purest silence,
Time suspended there
Looking on the timeless scene.

Lovers in their trance,
Death the deeper trance.
Dream of bliss, dream of death.
The strength of day wanes,
The blood of life pours out.

DESPAIR

O the snows last so long
And O man is such a brute
And O the girl is snatched by death
So fast you would not believe it,

And O I wanted to be tall,
And O the unendurable cold,
O that my mother was put away,
The long, the long trials,

And the love I had for one
It all went to pieces,
That girl who was murdered
And O that other my friend

Who set herself afire for nothing.
And O I hoped to be well
I wanted to be serene
O the snows last so long.

SUICIDE NOTE

I take no virtue of this as I finger the hand gun,
Insulted as I am by too much power of life,
Too small, too hurt in an alien place too large
To manage belief, victim of abominations.

Beyond the horror and the terror of the grave
I can make no moral distinction of good;
Evil overwhelmed me. I require
Obliteration and shall give it to myself

With one pull, impersonal, final, and timeless.
How to cross back through the vile tracks
That have brought me to this sane conviction
I am too depressed to know, too weak to care.

Veils and mists, mists, fogs and veils
Cover my life from the beginning to this end.
The tricky ways of the living were without help,
I go useless to the useless land of the dead.

It was the unrighteousness of the bad Germany
Without the hope of kindliness or mercy
Has brought me to the edge of the peace of my ruin.
I attest I cancel the letter you sent me, God.

EVENING BIRD SONG

When the birds are singing in the bushes
They harp on evening.
I sing with them beyond disaster.
This generation does not know
The extermination camps of World War II.
This generation of birds does not know
One generation ago, its own generation.
How they inflate the exuberance of nature!
It is evening and all their song,
Pervading the pine trees in full summertime,
Reduces me to mindless ecstasy,
Increases me to nature's impersonality.

Nature careless of the human condition,
Hear the birds pour forth their vitality
In purity of existence beyond disaster.
They will be doing it a thousand years from now.
It is evening. Darkness comes. And bird song.

David sang, and lived, in part.

THE SECRET HEART

In the secret heart all men are free,
Each reaches for himself in his secret heart.
Try as we may for social harmony
Conflict strikes us in the morning,
Time takes us away in its mystery.

Night reveals the essence of our predicament.
Relinquishing consciousness, we sleep as dead
And arise to face eternal quandaries
Gripped by nature as was the first man,
Struggling like a medieval mendicant.

The status of power prevails for a while
But we sense in the health of our state
The sinister destroyer in the undiseased eye.
Eventually all ideas go underground,
The only triumph is some elegance of style.

To have lived a few decades virile
Is to have seen, say, in the city of New York
No sensitivity to the saving of monuments
But a race to destroy the historic part
As if the past were totally sterile.

Think, all living now will in time be dead.
Not a breather now who dreams on eternity
But will have it thrust on him early and surely.
We are compelled in this great adversity.
Some few will some great truth have said.

In the secret heart all are free,
Each has a secret heart of pure imagination.
Each man struggles with reality.
In the dark scales of birth who shall not say
That each man in his secret heart is free?

TIME PASSES

All is a kind of toys, all is a kind of play
In the great stack house of poetry.
Whether to say it out, or play it fey
Deposits truth upon society.

Which does not know which way it went
Until shufflings of many a fall
Settle the account of each event
And show at last that style is all.

The toys of the mind, the toys of the word
In high displays, in richest glows
Tell it better than it was
When the shifting heart would come and go,

So that society knows now
In the great stack house of poetry
The soul that shone upon the snow
And the eye of every pain.

BROKEN WING THEORY

The poet comes with his white pain to save mankind.
Few heed him. Necessity grips the neck and foot of each.
I see them pouring through the streets of years and decades
Bent on the tasks assigned them by necessity.

Most in words, most are wordless.
They do not think through, and if they thought
The spectre of death would arise to throw them down.
The poet comes with his broken wing to teach them flight.

THE FISHER CAT

Wildness sleeps upon the mountain
And then it wakes in an animal
And in us, and in the sophistication of city streets
And in the danger of the indifferent murder,
We see the fisher cat on the limb of the tree,
Or is it a marten, or what is this slim, fierce beast
Caught in the flashlight's glare at night in Vermont,
Ready to leap at the baying dogs?

This enemy, this ancient foe, what is he?
The unexpected beast glares down from the high branch
Ready to pounce and fills man with fear,
Some nameless fear of millions of years ago in the forest,
Or the Rift valley in East Africa
When it was life or death in an instant.

The man has a gun, the instrument that has saved him,
Without which the drama of this intense moment
Might have ended in the death of man, and no poem,
He raised his piece like a violator of nature
And aiming at the jeweled caskets of the eyes
Brought the treasure trove of brain and sport.

The beast fell to the ground, unable to comment,
His beauty despoiled that took millions of years to grow.
The dogs thrashed around a while and quieted.
The New England hunter then with his matter of fact,
Taking for granted his situation mastery, put the
Mythic beast in the back of his four wheel drive vehicle.

It has taken the scientists of the university months
To decide what kind of an animal the creature was.
The centimeters of the back molars were counted,
Books were consulted, in the end it was decided,
Not by the scientist but by the poet, that a god
Had descended on man, and had to be killed.

314

Wildness sleeps upon the mountain
And when it wakes in us
There is a perilous moment of stasis
When savagery meets equal savagery.
The long arm of man maintains intelligence
By death: his gun rang out instant doom.
The paws of the animal were very wide,
The claws of the beast were wide, long his thrashing tail.

READING ROOM, THE NEW YORK PUBLIC LIBRARY

In the reading room in the New York Public Library
All sorts of souls were bent over silence reading the past,
Or the present, or maybe it was the future, persons
Devoted to silence and the flowering of the imagination,
When all of a sudden I saw my love,
She was a faun with light steps and brilliant eye
And she came walking among the tables and rows of persons,

Straight from the forest to the center of New York,
And nobody noticed, or raised an eyelash.
These were fixed on imaginary splendours of the past,
Or of the present, or maybe of the future, maybe
Something as seductive as the aquiline nose
Of Eleanor of Aquitaine, or Cleopatra's wrist-locket in Egypt,
Or maybe they were thinking of Juliana of Norwich.

The people of this world pay no attention to the fauns
Whether of this world or of another, but there she was,
All gaudy pelt, and sleek, gracefully moving,
Her amber eye was bright among the porticoes,
Her delicate ears were raised to hear of love,
Her lips had the appearance of green grass
About to be trodden, and her shanks were smooth and sleek.

Everybody was in the splendour of his imagination,
Nobody paid any attention to this splendour
Appearing in the New York Public Library,
Their eyes were on China, India, Arabia, or the Balearics,
While my faun was walking among the tables and eyes
Inventing their world of life, invisible and light,
In silence and sweet temper, loving the world.

MEANINGLESS POEM

If I could write a poem without meaning,
To escape meaning altogether, to escape it,
For meaning ends in suffering. Oedipus,
Lear, Hamlet haunt me. I knew them young;
Now I know them still. They mean so much
Time does not lessen their meaning
But bears it harder to the heart as age increases.
As age increases the age decreases, time blurs
The focus hopes had on life's meaning.
The genius of man stated suffering.

The poem I would write without meaning
Would be the poem in the heart of every man
When he realizes his own obliteration, accepts
The senselessness of existence, knows he cannot
Stop the bird on the bough from singing, cannot
Know or control the destiny of his nation, himself,
Or of man; understands the disparities of reality
And of imagination; confronts the limits of will;
And as time forces him toward nescience,
Understands the indifference of the world.

But no, no poem can be written without meaning.
Words express a desire beyond suffering,
Worlds of religious significance suffusing
Illimitable reaches which we dimly perceive,
As when fog unwilled lifts, illuminating us.
It is a keen situation with our suffering
When we cannot accept suffering as totality.
The inner self sees a flash of eternity
As fog lifts, and knows in wordless immanence
Goat-footed gods splashing light.

HOMAGE TO THE NORTH

After the swirling snowfall, days of whole
Downfall of individualized snowflakes' major patterns
When we work to keep up with the inches falling,
The world becomes a heavy lightness of light brightness,
One is given over to the softness of the December snowfall
As a foot or maybe an indeterminable two or three feet
Are swirling down from the heavens as time passes.
One feels the mass of it, wonders about danger
For man is rather quickly all too small a creature
Exulting in the primordial power of nature which may
Turn him to the opposite of his expectations,

And then the great cold like a clear white sting
Pervades the atmosphere as the snows cease to fall,
It is the power of the absolute god of winter.
We hover inside houses as night turns deadly cold
And awake to the awe and chance of thirty below zero.
Let man exult if he can, and he does, but narrowly,
We walk abroad in a kind of bravery of danger,
The invisible spectre of death visible before us,
Walking in cold that can kill, in our foot-salutes
To the white virtue, and in our heavy wraps
Clenching dictates from the god of the North.

AS IF YOU HAD NEVER BEEN

When I see your picture in its frame,
A strait jacket, pity rises in me,
And stronger than pity, revulsion.
　It is as if you had never been.

Nobody in the world can know your love,
You are strapped to the nothingness of ages,
Nobody can will you into life,
　It is as if you had never been.

I cannot break your anonymity,
The absolute has imprisoned you,
Most sentient, most prescient, most near.
　It is as if you had never been.

THE BREATHLESS

Something breathless hangs in the air
Of Gallinas Canyon, under the distant mountains
As if every dream of man suspended
Were as real unrealizable
As dreams realized in the here and now.

As that the brook is musical under Tres Alamos,
That the colt is elegant, in an absolute frisk,
That the Penitentes will never fail in their mission,
The spirit of the hermit is on Hermit's Peak,
That nature is ever a mystery to man.

If I could isolate a single quality
Of the perturbations and ravages of time,
Deny momently the forces of destruction
We inherit and shall not break, let me celebrate
A breathless quality in high mountain air. It is hope.

STEALTH AND SUBTLETIES OF GROWTH

I thought I could leave my habitation for the summer
Without the stealth and subtleties of growth
But when I returned to the summerhouse in Autumn
Vines and tendrils had shown their accumulated worth.

As if to say, you are dying, but we are living.
You go away to sea to champion new horizons,
Inveterate of lust, man of hope for the new.
You should have stayed home. We are your natural expression.

We have reached beyond the pruned and narrow.
We are the victorious life of vegetation.
We covered over the splendours of Angkor Wat,
We rode down the pyramids of Tehuantepec.

We may retreat from your tread and your private sickle,
It may seem as if you had commandeered our life,
Yet we, like you, look for the unwary principle,
March our fecundity as an absolute.

I almost heard the sermon of the tendrils
When I entered the airy summerhouse in September.
With deft expertise of strength of finger
I bent the vine, I broke it, and had some peace.

While you were away in the fields of imagination,
Cropping that hay, we have been gaining.
You may think you put us down with a finger
But we shall arise beyond your far dimension.

You are a twin of nature, but time will lop you
After millennia. You tried to conquer nature.
Too much intellect begat your downfall
In wastes of space. We do not outdo ourselves.

I saw the voiceless tenacity of green shoots
Overwhelming the earth by rigor of law,
Then laying an admonishing finger to a tendril
Murmured, "I live, you have no imagination."

EMILY DICKINSON

He saw a laughing girl
And she said to him
I must take a man
Toward eternity.

Her flesh was soft and fleet,
Her mouth was like a pose,
And a spiritual drift
Played about her flowing clothes.

She said she could not be
An evidence of the free
Unless she left her body
To become immortality.

She took him in the main
And held him in a trance
Who never knew for thirty years
Whether she was the dancer or the dance.

II

She was the highest mark
To which he set a snare.
He held her in his clutch
But she vanished in the air.

She departed with the years
And rode upon her destiny
While he was retained much
In the hold of mystery.

Now he must forswear
The roil of reality
And must admit the truth
Of what he cannot see.

She is gone with the wind
And he is gone with the weather.
Only in spirituality
Can they be said to be together.

Pretend to the flesh,
But the flesh will fall away.
In timeless uselessness
Love can have a stay.

He thought he held her
When passion was high.
Time brings her to him
In a long, in a wind-drawn sigh.

HARDY PERENNIAL

In youth we dream of death,
In age we dream of life.

I could not have cared less for life
When young, employing savage pursuit
Into the glories of the unknown,
Fascinated by death's kingdom.

The paradox was my brimming blood.
My bright, my brimming blood, my force
And power like a bridge to the future,
Could not contain itself in white flesh.

In youth we dream of death,
In age we dream of life.

Now that death's savagery appears,
Each day nipping at my generation,
The hard facts of the world negate
Symbols of the mind striving otherwhere.

I would give love to every being alive,
Penetrating the secrets of the living,
Discovering subtleties and profundities in
Any slightest gesture, or delicate glance.

QUARREL WITH A CLOUD

You rise like a mountain over the white building.
You are white, surrounded by blue.
You change me as I change you.

My quarrel is insurmountable.
You alter the landscape and change
Your being even while I am looking.

I have to think that I change you,
Being a man of will, but you
Do not know what you are doing to me.

It is an ancient dialogue
Of the inner and the outer world. I
Cannot, after all, contain you.

Already you have soaked up color,
Blushing rose; amorphous cloud,
You are undoing yourself among high airs.

The white building seems little changed.
You are vanishing in the heavens
And I am struggling with the word.

If I could fix you on a page
It would be an arbitrary election.
Clouds come and go. So do poems

In their world of white, and blue,
And rose, a make of the world,
Apparent, yet unable not to change.

Now I lay down my quarrel
In the heart of the word, rejoicing
In the nature of visionary change.

323

GNAT ON MY PAPER

He has two antennae,
They search back and forth,
Left and right, up and down.

He has four feet,
He is exploring what I write now.

This is a living being,
Is this a living poem?

His life is a quarter of an inch.
I could crack him any moment now.

Now I see he has two more feet,
Almost too delicate to examine.

He is still sitting on this paper,
An inch away from An.

Does he know who I am,
Does he know the importance of man?

He does not know or sense me,
His antennae are still sensing.

I wonder if he knows it is June,
The world in its sensual height?

How absurd to think
That he never thought of Plato.

He is satisfied to sit on this paper,
For some reason he has not flown away.

Small creature, gnat on my paper,
Too slight to be given a thought,

I salute you as the evanescent,
I play with you in my depth.

What, still here? Still evanescent?
You are my truth, that vanishes.

Now I put down this paper,
He has flown into the infinite.
He could not say it.

THE TRUNCATED BIRD

He stepped out of bed at night.
He turned on the light. In the blur
Of the first, stupefying sight
The upper half of a scarlet tanager

Lay on the rug. The severed bird
Lay like a comic mask, dropped
By the cat in the night unheard.
For a moment everything stopped.

Equal to supposing life whole
He lived in a forceful equipoise.
Some say this was the soul
Burgeoning while it subtly destroys.

MAN'S TYPE

When he considered his linguistic fallacy
He was thrown back to the primitive subhuman,
In consternation at the rise of man,
English not lasting millennially.

Something attractive in that slight figure
In the Rift valley millions of years ago,
Slinging his weight, craft outwitting his prey.
The crudest action would be long to stay.

Deft we are still to maim and kill,
We have the big means, the lack of sensitivity,
The annihilating energy.
Of redeeming grace who shall say?

LONG TERM SUFFERING

There will be no examination in Long Term Suffering,
The course will come to an end as planned.
I have found that examinations are useless,
We have altogether too short a time to spend.

Time, ladies and gentlemen, is the great examiner.
I have discovered that this is true.
It is what you write as you go through the course
Is the only determinant and determinator of you.

Long Term Suffering is for those of all ages
In our tussling University, our bulging classroom.
It may be that I will profess near madness,
It may be that you will write out your doom.

All that you will have at the end of the course
Is writings you indite, or poems you make,
If you make them. Words, words in a sea flow;
At any rate, a lot of heartbreak.

Save your papers. It may be that years later,
Forty, maybe, you would like to look back
At your course in Long Term Suffering,
And note how strangely you had to act.

YOU THINK THEY ARE PERMANENT BUT THEY PASS

You think they are permanent but they pass
And only contemplation serves to save their memory.
You are in the Pan-World building with the leaders.
They all seem real, they all seem permanent

But soon you are in the Pan-World building with the past.
How closely and with what immediacy
You scrutinize the features of each noted face,
President, old poet, Supreme Court chief justice,

Secretary to the United Nations,
How lively their speech, lively their looks,
As all together in one banquet place
You would think there would be no end to this.

Who in the reality of his high days
Thinks of the destitutions of the night?
You think they are permanent but they pass
To feed ravens of the ravenous past.

HATRED OF THE OLD RIVER

I come back to the old river
And it is still going through the hills.
The seasons change, the years leave,
But it is still going through the hills.

My exasperation is exquisite and complete,
It is ineffectual and pathetic.
I cannot change the course of the river.
I cannot draw anything out of the banks.

The river does not care how I am.
It does not show time's worsening,
Seems to look not a year older.
That downward course is going against me.

I come back to the old river
Savage against its indifference.
Time takes the savage out of me.
Time is taking me out to sea.

It is as if I confronted my parents
At the end of life, passionate,
Asked them momentous questions
But, obliged to look upon their tombstones,

I cried out to God the Father Himself,
Giving credit to the supernatural,
Expecting from it a natural answer,
But receive primordial silence.

The ancient rebuff of the spirit,
Or if there is a sound at midnight
It is an alarm so deep
That for an alive mind there is no sleep.

VERMONT IDYLL

These are the days of yellow and red
Thrown up across a far field,
October's eyeball-striking glory,
A day that imitates the summer.
The leaves are falling, will come winter.

You lie upon the grass, the sun is hot,
Your skin is moist, you think of summer,
But when you stand and walk a cool
Cleanliness of the lengthening day
Reaches a white winter in the bones.

And then the silence of this time
Is opened by an engine's oncome
Coming slowly with growing invasion,
Changing meditation,
Until goes past, across, the manure spreader.

The red and yellow sentinels stand by.
The lurcher machine, immigrant, throws out
Its rich burden over unplowed grass,
A secret ritual or rebirth,
The hand of man applying the levers.

Only a moment without man's agency
It seemed a timeless perfection
Was one with consciousness,
A stasis like a dreaming mind
Between summer and winter.

Nearby, a car rotting among thistles
Had jagged glass, teeth of broken windows.
In the back, a ruined cushion with a hole
In the center. A discarded plow rusted too,
As time was stalking.

THE SCOURING

A stubborn man put out to sea
From Bucks Harbor to the Bahamas,
Two couples in a handsome vessel,
A storm came up off Connecticut,

They turned back to the north and east,
He thought he was master of the sea,
The night came down, the wind came up,
Instead of putting in for Portland Light

They aimed the boat for Monhegan Island,
No riding her out far out,
No setting of a sea anchor,
They missed the channel between the rocks,

In a wild night of fog and storm
Met death's scour on the southwest rocks,
The rugged rocks of Monhegan Island,
Four bodies in life jackets found in the morning.

Four are dead who might have lived
If sense had sent them prudence,
But for a primitive dash of prudence,
If they had only had a little prudence.

THE CAGE

Emily and Kate
Were two birds in a cage.
One got out,
One stayed in.
If mismated,
Both were fated.

330

One was lost in the world.
The other in her cage
Pouted, preened, looked inner,
Became white and thinner,
Tall, secret
In electric rage.

Kate and Emily
Were two birds in a cage.
One got out,
One stayed in,
Tamped herself to death,
Stamp, stamp.

THE POET

He hopes his arm is strong as the gods'.
Desperate confrontation of the godhead
Leaves him jumping in a comic attitude.
He is the world's jack, jumping jack,
Jack Nobody, Jack Hope, Jack Kennedy.

How many sufferings do we have to suffer
Before we accept life as suffering?
Nobody believes it. O the belief in joy!
I jump at joy, highjacked by joy.
I cannot believe that mankind is here to die.

Death, an abstraction of the intellect.
The joys of the flesh are quick,
The dark spirit is quick, to open Lear is
To sense fullness, total belief
In endurance inheres in the muscle of man.

331

Down death! Down, Donnean absolutist!
Set down daring life to the death,
Joy of creation, illimitable illusion.
Put aside strides of thinking,
Election: red realities of feeling.

MAN AND NATURE

Man is the writing instrument
Who thinks that he impresses nature.
He perceives the sense of the sea
As if for the first time.
It was so decades ago.
He sits as a contemplator
In secret frenzies of realization,
The islands are the same,
In the spirit of indifference.

He leans forward in his seat.
He has faced the headlands.
It is a place too pure to be true.
The tensions of his spirit leap
Looking into ultimate ocean
Beyond Spectacle, Pond, and Hog.
There on the horizon Eagle Island
Light and the open Atlantic
Deceive him again with thought.

The summer trees, the birds, the bees
Eradicate fate of timelessness
In high July, but not for long.
The long thoughts are beyond.
As he looks outward to the oceanic,
Beyond a child coming in the door,
Beyond the self, beyond desire,
He breathes the universe,
In, continuance, out, salutations.

OLD TREE BY THE PENOBSCOT

There is an old pine tree facing Penobscot Bay,
On the bank above the tide,
Like a predecessor. Its fate
Will be that of its forebear.
I watched the former tree ten years
While it faced the surge of the sea.

Whelmings of the tide, line storms
Buffeted the root system of the pine
Twenty feet above the tide.
I watched the changes of the seasons,
Each year returning assessed
Somber change, a kind of stalwart declination.

As children grew up the pine tree grew down,
Threw down its length in defiant slowness,
Until one summer it was almost horizontal.
Even then, with jags of dead branches, it clung
To life, until further summers dipped it down
Until it lay a dry myth along the shore.

There is something ominous in the new tree
Erect on the high bank, at the very edge,
The sea's hands pulling out earth from the roots,
Slowly displacing a system of boulders,
As I watch through soft afternoons.
The tides have slowly taken the children away.

Then this new tree, about to begin to fall
Through subtle gradations of the strength of years,
Took on the force of a grisly apparition,
My memory was forced down to defeat,
My riches gone, my corpse on the beach,
Dry bones, dry branches, I too a myth of time.

PLACATION OF REALITY

Clear the music, clear the message, the mind clear.
The year is falling again, as it ever fell.
I put on Handel and put off cacophony,
The times are changing in a blaze of violence,
The soul is weak and the heart is brutal,
And evil is everywhere in the nations rocking.

A purity of the ancient soul
Which forgives evil before it has happened,
Understands errors of man before he makes them,
Sees a united vision in the skies
Where the head and the heart are sealed in a vision
Rugged as the structure of the universe,

A primitive idea of the whole man,
Whose words are of the soul, and with a pure word
I will clench chaos, and I will dream truth.
Old longings, beyond the word.
The year is falling again, as it ever fell.
Clear the music, clear the message, the mind clear.

EMBLEM

A great snowy owl
Sat on a cupola
Looking wise,
But undoubtedly thinking of mice
Who might be down under the barn
To come out to be devoured.

He had flown in from the far north,
So large and commanding a bird
Hundreds came daily to gape.
His head swiveled,
His eyes were astonishing,
His silence complete.

Where is the snowy owl of yesteryear?
Overlooking the snowy plain,
The snow hills from his perch
Atop the cupola,
Gazing at small creatures
In and out of cars below.

Where has he gone, honorer?
The cupola is unadorned,
The site as it used to be.
In deprivation of eye joy
And nature dressed to fineness,
My heart flew north, absolutely white.

WORLDLY FAILURE

I looked into the eyes of Robert Frost
Once, and they were unnaturally deep.
Set far back in the skull, as far back in the earth.
An oblique glance made them look even deeper.

He stood inside the door on Brewster Street,
Looking out. I proffered him an invitation.
We went on talking for an hour and a half.
To accept or not to accept was his question,

Whether he wanted to meet another poet;
He erred in sensing some intangible slight.
Hard for him to make a democratic leap.
To be a natural poet you have to be unnaturally deep.

While he was talking he was looking out,
But stayed in, sagacity better indoors.
He became a metaphor for inner devastation,
Too scared to accept my invitation.

A MAN WHO WAS BLOWN DOWN
BY THE WIND

There was a man who was blown down by the wind
Who had a sense of adventure, desired the moon,
Might have been bright Apollo in a dark place,
But it was enough of an adventure
To ride a parachute in a Colorado,
Cloudless, frenzied and quaking
Sun-bright windstorm.

The parachute instructor had two men with guidelines
To steady the craft. The protagonist
Requested photographs to record his gallant
Flight heavenward, assertive, imaginative.
The young man implicated himself in the harness,
Fixed under the big chute, nearly
Horizontal in the stiff wind.

He gave a signal of ascent. Here language halts
For nothing else was said, and nothing can be said.
The arrogance and tragedy of man is reasserted,
Once more his defiance is demonstrated,
The wind in the clear day, without a cloud,
Was rushing down from the Rockies, tree-
Bending at eighty miles an hour.

It registered one hundred and twenty on the peaks.
Here it was bowling stiff, and hard, and bright.
The strapped man took off in his high flyer,
Leaping up from helpers, looking like a god,
Apollo-bright boy, dreamed of by boys.
He will fly with faultless daring to heaven,
Self-designated as the able master of nature.

The flight of the soaring man was photographed,
Expert-clicked. Inexplicably, but who would not
Have thought it, the winds frenzied at eighty,
A hazard stroke, and down-draft whiplash
Dashed our believer instantly to the ground.
When he hit the ground he was upstanding,
Out of luck, accordioned and killed on impact.

I never knew this man, neither did you,
You never will. I see him riding the kite of hope
King of the comic strip, batman, superman.
A poet should acclaim his climb to height.
What is the correct attitude to take?
No attitude will shake the fall of man,
Not Apollo's, nor that of Icarus.

THE HOP-TOAD

The hop-toad jumped away, missing the blade,
When Betty was mowing the garden.
She instinctively said, I beg your pardon.

UNITED 555

St. Paul never saw a sight like this.
Seven miles up, fifty five below outside,
Wide open flat top of cloud vista to the far horizon,
As the sun descends reddening the upworld spectacle.

St. Paul never got off the ground, and, for that matter,
Christ was nailed to a Cross a few feet above the earth.
Here I sit seven miles up feeling nothing,
No visceral reaction, dollar martini, endless vista.

I must say it could not be more beautiful.
O think of Akhnaton, who never got off the ground either.
Raciest to think of the baboons in East Africa at Treetops,
Who could not imagine to come to such a pass as this.

Christ and Paul never knew what height is,
They never polluted the atmosphere.
I am Twentieth Century Man riding high,
Going into the sunset, Seven Up, feeling no pain.

LIGHT, TIME, DARK

LIGHT
I approach upon the headlands
Impressed with my own beneficence.
I come early, opening my hands
To show world magnificence.

TIME
None has subtlety but me.
I am the only one who is free.
Because I am adamant I give
The mystery of the relative.

DARK
I persuade with the impenetrable.
My power is whole and able.
Lost is the human face
In the depths of my embrace.

TIME
Each makes his discursive claim
Reassured in his individual name.
The light would treat of the light.
The night treats of the night.

LIGHT
See the children playing by the shore
As if they had never lived before,
Their heavenly faces shining
With my divining and defining.

DARK
See the old robed in doubt
Where light is all but put out
As I draw them to the senseless
Depths against which they are defenseless.

TIME
I am ever the master here
Between man's joy and his fear.
None escapes me and all shall
Bow down to the impersonal.

LIGHT
I have the power and the glory
To tell man's story.

DARK
Beyond action and thought I grow
Beyond what it is to know.

TIME
Mine is the true delight
Penetrating both day and night.
Mine is the mystery
Behind everything man can see,

The subtlety
When every man alone,
Dreaming that he is free,
Knows me in the bone.

DEATH IN THE MINES

Think of man praying. He raises his hands to God.
Whatever his doubts, he has come to this attitude.
It is a skyward and an outward penetration.
The man praying tries to penetrate mysteries which are heavy,
Which turned his hands downward to earth, its common work,
In any hope beyond the common disasters of time.

Then I read of the miners of Cape Rosier
Who, descending into the interior of the earth,
Exercised their hands upward to pick the rocks above,
As if they could uncover and pluck some ultimate stone.
Thinking not of Samson pulling down his temple,
They struck (one stroke!) one ore so rich in meaning
Devastation shook, and killed them in a pile of rubble.

They are dead, a common lot. America goes on.
The nation rides on the skin of the planet, multitudinously.
As malfunctioned astronauts might ride around the planet
Dead until they disintegrate to a cinder, a puff
Finally as light and delicate as an April daffodil,
Influential members of mysterious time,
The dead miners in the slow growth of their disintegration

Express the serious, interior reality of poetry.
Ride easy, earth, in your strong contention,
You are stronger than man, and ride us down.
But a wind will spring up, a spirit arise
And ride on the air lightly, supremely clear,
In other centuries, and in other civilizations.

ADAM CAST FORTH

Translated from Jorge Luis Borges

Was there a garden or was the garden a dream?
I ask myself, slowly in the evening light,
Almost for consolation, without delight,
If that past was real or if it only seems
Real to me now in misery, an illusion?
   No more than a magical show
   Of a god I do not know
But dreamed, and that Paradise, vague now, delusion?
   But I know that Paradise will be
   Even if it does not exist for me.
The warring incest of Cains and Abels is the tough earth's way
Of punishing me. Yet it is a good thing to have known of
Happiness and to have touched love,
The living garden, even if only for a day.

REDEMPTION

I am a profound man sitting in a park.
Beside me is a wren sitting in the dark.
Up in the sky is a god planing on a cloud.
He sings aloud.

If the wren were on the cloud
And the god were in the park
And I were in the dark
Who would be proud?

Now I am a man making shields
And there is a woman wearing white.
A black slave is rising
Whose joy is complete.

If I were a man without a shield
If the woman were always wearing red
If the black would sulk
How should I live?

Each stone I turned began to sing
In the park of wheeling lights.
What it was, was not the same again,
I joyed to see the ego plain

And turned to stand and stay,
When bird, and god, and man
And woman emerged as their true selves
That put my mind away.

UNDERCLIFF EVENING

I feel illimitable essence,
As if I could express everything
Known to man. This omnipotence
Is a gift of nature in radiant acclaim

That I can know a secret of being.
This instant a parti-colored butterfly
Flew across my sight, real as evanescence,
Startled me in its non-verbal reality.

Bandied butterfly, severed spectacle,
Coming to consciousness for an instant,
You are a messenger of supernal power.
You escape the words of poetry.

I try to put you in words, you are evidence
Of a dream of insubstantiality,
A moment of perfection in the dream of time,
You vanish forever, I shall not know you

But as a feeling of illimitable essence,
When the world, despite error and chaos,
Announces gripped summits, elects to show
Mysteries beyond our tongue, shafts of light.

I did not ask you to come to me,
I did not ask you to fly by my face,
But you erase my heavy thoughts,
You bring me into lightness and grace.

PORTRAIT OF RILKE

I saw a picture of Rilke
Leaning in a doorway, tentative,
Pausing.

I was young and he was near
The angels,
The angels he loved in his long lines.

I never thought so long from then,
At least half a life-time,
His gesture would return,

An absolute datum of reality.
So slight a gesture
To mean so much!

I see him leaning in a doorway,
In a slight frame, with all he knew
Of life,

Lightning fixation in time,
Enigmatic, sublime,
Tentative! Tentative.

SPHINX

I saw the Sphinx on the sands of Egypt
Jealous of my love, jealous of my power to act,
Dead for centuries, completely out of the picture,

And I saw a sphinx who did not care
Whether men win or lose, whether women
Live or die, but I care and I live
And I love and I dare to say the truth,

My truth, this truth, this imperishable instant,
My unresolved questions better than your absolutes,
I would rather be alive, suffering and exulting
Than living like you in a stone tomb
For thousands of years in a silent enigma.

And I see Shelley plain, too plain, I see him
Sensitive to the exigencies of my own condition,
Trying to make peace with your eternal enigma,

But I reject you as a stone monument,
We are not stiff, we reject stone,
We believe in the fervour and felicity of the blood,

We believe, the young and the old, in immediacy,
The holy moment, the impact of the present,
A moment of grace, the love of the flesh

Greater than the lectures of history,
And we believe that a Sphinx cannot see
But we see, and we love and present

The evanescent as more durable than your stone,
The look of the eye greater than blindness,
Love of the flesh caught in love of the spirit,
Love of the flesh inevitably spiritual,
In love of the flesh is the most spirit,

So I say on the west coast of America
On another continent away from Egypt,
Sphinx, witness of time, we are your judge

And I force upon your stone my spirit
Inevitable as your unanswered question.
When answered, you killed yourself.

Oedipus drained his eyes of light,
Preferring the blindness of insight.
I accept man two-legged, or three,

Loving mystery, in which to rejoice,
Not averse to love, love's mystery,
Love is the mystery in which to rejoice.

THE GROUNDHOG REVISITING

He has gone over to new flowers in a garden
And deliberately deflowered precious aspects
Set out for the wedding of Gretchen,
And thus become the enemy of the family.
One cannot stand this, stand destruction
Of fair flowers for a field of folk
To wed man, woman and nature to the future.
We do not want to kill him,
But dissuade him and discourage him,
So I secure gasoline in a can
And pour it down his holes.
Get a drift of materialistic America.
On with marriage, down with the groundhogs.

I do not understand the animal kingdom
Any more than I understand the human.
Gretchen could turn six cartwheels
Outwitting my power to put on paper
Pure agility and grace in action.
She turned them, elegantly, on this lawn,
And now she stands here to be married.
It is another turn of fortune, may fortune bless
The turnings of their future life,
Grace this company in some retrospect
To remember, and think on youth, for
Men and women make glory of love,
We are here to celebrate love and belief,
May time bless these believers, love give them grace.

BIG ROCK

I have sat in this garden yearly
With an eye to the changes of nature
But nature goes on changing, wryly,
Indifferent to my attitude.

This excites me. I sit watching the sunset
But it was the same decades ago.
Nature is impervious to my bodily changes.
I am amazed to see trees, skies unchanged.

But what of five hundred years from now,
Or one thousand? Do I dare to think of this?
How will nature have fared by then?
I, long a part of her.

I claim that I make nature alive
Because if it were not for the human predicament
Nobody would know what nature was like.
I can specify particularities.

The Madison Boulder, largest erratic boulder
On earth, was set down by the glacier
Before man dared walk on New Hampshire;
Huge-tonned, rectangular.

I think of the Indians who moved around it,
I think of the early Americans,
I think of youthful lovers in the Spring,
Old couples contemplating philosophies.

I can imagine anything I can imagine,
Nature is not my lover, I am the lover of nature,
I kick the boulder with the foot of Hercules,
The boulder sits, but I can walk around it.

AMERICAN HAKLUYT

Go to the depths of willing
  Sailing in a boat
  Which argues
  Staying afloat

And you will see a destination
  Like a nation
  By navigation
  Making for harbor

You willed to go sailing
  Because you were bright
  The waves enticing
  Until night

It was not as if seas
  Controlled your method
  Nor would block
  Your adventure

Life all a kind of gain
        You think you win
        A certain daring
        An interim

But oceans rise up
        And time closes down
        Agony comes
        And you drown

The nation falls on rocks
        The vessel never found
        Old historians
        Deplete Rome

LIFE AND DEATH

Jean Garrigue (1913–1972)

The agent death assails me now,
Your sufferings blanked out in the casket,
Your voice rises over the liturgy
Pure, unassailable, and final.
Your single voice of all who lived
Speaking the lyric that should live,
What are my tears coursed for you
Before the clear spirit of your poetry?

Dear Jean, and I went to the graveyard,
Proffered an ordered aesthetic,
Brilliant snowfield under a winter sun
Ablaze for a moment upon your flowers,
An incised chill, pure, transparent,
The scene sensational, crystal,
And in ignorance of the new you
All who knew you shared difficult grace,
The refusal to admit death,

We knew it would be ours too,
Who live to love you as living,
Charged by the clarity of the scene,
Unanswerable enigma present in the flowers
Made radiant by iced winter sun
In acres none would believe,
Then when all got into the waiting cars
The sun in cold winter snow glaze
Was overshadowed; and each in his depth
Left the second scene, still human,
As something intolerable,
Praising the voice of the poet.

FLOW OF THOUGHT

I would not despoil this page
With anything but the truth,
Yet the truth is elusive
And I knew it most in rage.

For it was by an assault
That I dared to state
The simplest realities of nature,
A trillium on the brink.

And it was because of suffering
That I decided to excise
Suffering, and speak of nature,
Indifferent to our suffering.

And it was because of love,
The love of endless opposites,
That I voted against suicide,
And lived for the love of man.

And it was because of love
That I caught out of the air
The fragile, frail
Sensations, and thought them immaterial,

And thought them immortal,
And hopes that we would see
Beyond materialism spirituality,
Grandeur possible, grandeur in man.

MIND AND NATURE

The characteristic of the mind
Is volatility, while that of the trees
(O will the trees not ever stop not talking)
Is to look elegant through seasons
Cold or hot, winter or summer, they stand
Grand, they stand taking the breeze, pines
I am thinking of, I cannot believe

Their body changes as my body changes
With time, but I have to admit
These tall pines have grown over twenty feet
In twenty years. Yet how indifferent!

Sensitivity of vegetables is the new teaching,
To love a plant is to get back its loving,
While if you hate it it will shrivel,
Yet in all the love of looking I gave the
Tall pines, I sensed their superiority.
They sway back and forth when called to
Without care for the human predicament.

A hundred feet in the air! A hundred years growing!
What do they care what happens to me,
They have never heard of Beddoes, or of Clare,
They reject one or other theory of poetry.

The characteristic of mind is volatility,
Whenever I see a rock face,
Majestic, standing high in the air,
Making one think of aeons
When force thrust up these harsh stones,
I think of the volatility of mind,
The shortness of earth duration.

No resolution! The mind
Is a play of lightning, a happy chance,
While the duration of the earth, fixed,
Makes us one with the bee, the ant.

WILD LIFE AND TAMED LIFE

(*For Ian and Trekkie Parsons*)

Man, tamed by wine, by dress, by talk
Looks over civilized acres, conscious
Of history. The Norman conquest
Was not such a long time ago.
Geologic time looms graceless before,

But it is the slow twilight of summer
As we sit inside looking out
For the great show. A fox will show!
Coaxed by scraps on a garden table
The sinuous vixen will come, jump

Up, seize food, jump down, slink
To side shadows, her male cubs too,
As furtive, daring, exquisite,
Will bring wildness to the scene
Of civilized man smoking in the parlor.

It is better than a Watergate TV show
To see the actual animals in play
In place, for food not crime, induced
By man the master to come forward
And be the example of our native state.

They live at an uneasy distance
From one another, fox and man,
Surviving centuries. Aesthetic delight
Claims our attention at this sight
Without articulation of our sameness.

INCHIQUIN LAKE, PENOBSCOT BAY

(To Jack and Moira Sweeney)

Galway Jack, Dublin Moira, wild swans,
Flapping the lake water, going a way off,
To settle under the big burren brazoning over there,

You have set your house apart from wildness,
The world to view spectacular,
Scope of history in a picture window.

Within all is warmth, Paddy in the glass,
The fine elicitations of the mind
Spark again your old subtleties.

To have gone back to the old world,
To have escaped from brain-blasting America,
To be an overseer of ancient Ireland!

Here the fog comes in from the Atlantic
Pumpkin Island far in mist, half alight
In evening half fog, a heaven pageantry

Of strangest lights and shapes, a drama
Never to come to a conclusion, slow
Drifting occlusion, hundred mile view,

The stealthy nature takes over the land,
Wigwam Indians prepare to fight the white man,
Summer folk see TV Watergate.

FACE, OCEAN

Stephen Spender, I write about your face,
A work of God, about which you had nothing to do.
I saw it in Sussex in the summertime,
More extraordinary than it was forty years ago.
I could not have admired more that craggy
Eminence, yet with a gentleness, a tenderness,
Some massive statement of august nature
(It was July) of the male and female principle.

A face for the age! But what is the pictorial
Against the weight and mass of the whole being?
How do I dare to write about your face
Without stating your entire intelligence?
The intelligence may be books forgotten,
Or wrong turns taken in labyrinthian ways,
While the sight of a statuesque face
May turn us on to time's attitudinizing.

I have become an absolute thing of the sea
And loll, somewhat far out, with flippers.
When the second tide brings warmer water
To the shore, as a creature of the deep,
Guessing there will be no sharks daring here,
I propel my body through the sea without
Will, with nature's force, loll and turn,
How pleasant to leave the mind back on the shore.

THREE KIDS

Three kids running down the meadow,
Frisking, cavorting, at the heels of Gretchen,
Young eyes, blond hair running wild,
It was a picture of Spring in New Hampshire
In high May when blood was running wild,
Young goats, a young girl, triumphant.

If I lived a hundred years
No ink of mine from a passionate hand
Could communicate to you, dear reader,
Essence of ecstasy, this ecstatic sight
Of joy of life, limitless freedom,
As the girl and the young kids leaped and played.

As the young girl and the kids played and leaped,
I remarked the rectangular eyes of the goat.
The lively harmony of color and line,
Yet two weeks later the kid was dead of disease,
The other two were sold, the meadow never knew
Again so high a sight in the month of May.

TRYING TO HOLD IT ALL TOGETHER

Trying to hold it all together
Is like trying to hold back bad news.
The inevitable, the death of Auden.

Shocked to hear of his Autumn demise,
You cannot take it back and say,
I saw him in June, in London, he looked well,

I take him for granted like nature,
Assume him totally assimilable,
Assume he will go on being kind and generous.

We are faced with the hard facts of time,
Beyond any man. We cannot outface time,
Nothing can be done about the human condition.
O nothing can be done! Don't think it. Don't believe
Will will help us, or religion, his civilized stance,
A comic attitude, any saving grace,

We cannot hold it all together, the depth,
We cannot trick it out with word embroidery,
Time is the master of man, and we know it.

A WAY OUT

I

Time mocks but I would mock it,
Throw hurricane force against its devil,
Commanding it to stop. Flaunt a panorama
Of reality. Absolute. Here is the truth!
I want to look at the world as it is,
Look into the eye of a stopped eternity,
Believe in what I see, the great exemplars
Appearing on the stage of inner vision,
Elate, final, two radiant believers.

Here I see Buddha making the great denial,
Leaving his wife and child, having seen
A sick man, an old man, and a dead man,
Giving his life as a living sacrifice
To denial of appetite, dedicated to spirit.

And here I see that other master of mankind,
Jesus the subtle master of those who know,
Greater than Aristotle, to whom Plato
Had access, the daring revolutionist
Who knew that He would outlive Roman materialists.

These mocked the surly naturalism in man,
Said he was better, lit ways to lead him.
They covered the East and the West with light
And we in our faltering century have
To aid us blind Marx, Freud, blind Einstein;
Godlessness, fear, and relativity. Extinguish
These three their gross fanfares and bonfires,
While the serenity of Buddha and the fury of Christ
Give mankind examples of the way to go,
The ineffable, and the active means to know.

II

I walked on China, a young man, searching
The Buddha, I knew what he knew, but
A fierce Western passion drove me to Christ,
Exciting my membranes with His pure sacrifice.

The torment of His impossibility, the Nietzschean
Knowledge that there can be only one Christ,
Gave me the world-wrestle of two thousand years,
Doubt and belief warring in me to this day.

Also the impossibility of achieving Nirvana
Mismated me with the serene ideal of the East.
My warfare was that of rationality,
I could not abrogate my reason East or West.

Caught in this dilemma, I dreamed of time
And flung myself on the breast and body of nature.
Naturalism claimed me as day turned to night,
But I was struck twice by a blinding light.

III

Now when ice comes over the river,
And boys skate before the coming of the snow,
How pleasant it is to sit by the Franklin stove,
Braced for ice and snow and thirty below.

And if we can endure the cold, and skiis
Will take us turning down the trails,
And we can walk on snowshoes from the barn,
And drink the milk of goats new and warm,

We can live in nature as in our mother
Before we were born, and we can sense
That old death will give way to new life
As new mornings grow, Spring comes over the land.

INCIDENCE OF FLIGHT

I spring joy out of my rib cage
Like a flash of pigeons flying North
South here in Mississippi, Florida,

I insist on the aspiring eye,
Try as time does to cast it down,
Cast up the eye, birds their blue nature

Transfer through the air from the soul
Whole in its ambiguous essence
From one place to another

Without waste, we follow them
Ten times higher for their flight
Because we dream the same dream

Teeming in space out of our rib cage,
Age shall not deter us, nor walking stale
Flying, we are going up high in joy

On blue air, as if birds were the spirit,
Man was meant to walk, but meant to fly
Joyful as pigeons full of the grace of space

And if we have joy we have love
Above all else. Flare up, love in the heart
Part of flying, spring joy in Mississippi, in Florida,

It is a tale bold as an ideogram,
Conclamant wings heading North
Forth away. Joy uncages man to love.

SLOW BOAT RIDE

The boat moves slowly through the fens,
Britons going three miles an hour to Ely.
If there are locks, they lock three feet of water.
A narrow gauge to gauge the time of the world,
As the thoughtful, the pensive, the sad, the gray
Hold onto themselves looking for swans.
By the side of the banks, as we progress,
There are somber men of indeterminate age
Sitting in dilapidated chairs, holding poles,
With no expression, although it is high July,
Looking for a strike, although it is unlikely;
Green slime fishes finicky to finger.
It is obvious, to me at least, they look
Into the water to look into themselves.
They look so darkly into the brightening day
It seems that they were here long ago.
They are statues of a thousand years ago
When these ancient men, pondering imponderables,
Fishing, fishless, to philosophy allied,
Heard the cattle munch, saw the birds fly over,
Stone on stone slowly amassing Ely cathedral,
Which duly rises to our sight across the fens,
The destination of an ancient time, while now
These modern dreamers, agloom, pipe-smoking, ghostly sit.

THE POEM AS TRAJECTORY

If the poem is a trajectory
It has to go somewhere definite.

The trouble with the mind
Is that it shoots off in every direction.

Should the poem go North,
East, South, or West?

Should it go anywhere in between?
A trajectory could hit in the dark.

Should it pierce the heart of Confucius?
Is its target Mohamet?

Should it pierce Buddha's secret?
Should it hit Christ's wound?

I ask questions about the poem
Because it deals with reality,

An intractable substance, which, if hit,
May favor timelessness.

SNOW CASCADES

Snow cascades from the tall pines,
Large powder puffs
Puff down in soft lunges,

Descending to earth,
Subdued incidence,
A sort of soul of January.

They do not have words for it.
If they are wordless,
So, finally, are we.

It is as if they would say
What we say,
I mean what I say by being.

360

COAST OF MAINE

The flags are up again along the coast,
Gulls drop clams from a height onto the rocks,
The seas tend to be calm in July,
A swallow nests under our areaway,
It is high summer, the greatest days of the year,
Heat burgeoning the flowers, stones heating the tides,

This is peace, the indifference of nature, another year
Seeming the same as the year before,
The static ability of the world to endure.
There is Eagle Island twelve miles down the bay,
A mole has just dared to march over our garden,
The far islands seem changeless through decades.

Yet, think of the drama! Here am I,
One year older into inevitability,
The country torn in honor's toss-out,
What does nature care about the nature of man?
Three hundred years ago along this coast
The Europeans came to confront the Indians,

Yet the Ice Age shaped these shores millions
Of years ago, unimaginable upon our senses,
What do I say to the beneficent sun
Descending over the pine trees, the sun of our planet?
What does it care for the nature of man,
Its virile essence unassimilable?

Here come the hummingbirds, messengers
Of fragility, instantaneous as imagination,
How could they be so iridescent-evanescent,
Quick-darters, lovers of color, drinkers of nectar?
Do they remind us of a more spirited world
When everything was lithe, and quick, and visionary?

USURPER

Seizures of power,
Nobody gave it to you,
Wave-rhythm out of the trees,
Something unseen in the river,
A sense felt in the air,
It has you, and you know.

Your will demands the truth,
Seeks, seizes, secretes
And gives in new forms
Strains of received being,
Art momentarily controlling
Myths of mysteriousness.

Not to be supine! Not to lie
On the face of the earth dreaming
As a boy as you used to be,
Your eyes full of sensual hope,
But to stand up to nature,
To stand up to time, and say
I seize your power, I am.

VISION THROUGH TIMOTHY

Vision through timothy
Is different from vision as if it were clear.
Looking at reality, looking at the sea
Through timothy, occludes partially the view.
It is meshes of extreme nature,
You cannot see through nature's extremity.

You see partially. This is the reality of vision,
To see partially through nature's extremity.
Veils were drawn over our eyes
When we were born, or else
We would have knowledge of eternity
Upon our eyesight, yet we see

Only man in his adversity, man incomplete.
We set up great universities
To see man complete, yet his nature
Eludes us in our strenuous studies,
We cannot determine the reality,
We are caught in the nature of perplexity.

Love leaps in the nature
Of vision through timothy.
Timothy comes lifting in July,
July is a high time of the year,
If I were sitting here
In October, timothy would disappear

And I should be more bland, more austere.
I would think I would see
The world as it would inevitably appear,
But would I be more human,
In a vision so incontestable,
Than looking through timothy to see the sea?

ONCE MORE, O YE . . .

The grim New Englanders stand around
As in storybooks they stood,
Putting up with things as they are,
Another death of all the deaths they knew.

Our friend has departed, he did well,
His companions and townsmen stand around
As the minister intones the gospel.
They have no expression, all dry of eye.

They have no fear, and their belief
Is cold as the wind on the first of March.
They know the world will go on,
They do not raise their voice for a psalm.

The old strong characters of the North—
The doubting old Christians of Vermont.
Their faces immovable as marble
Used as monuments after such occasions.

RIFKIN MOVEMENT

Don't let it mean anything,
Hold back the meaning,
But let it flow.

Sweet, strong, and pure,
Your piano is your soul,
Let it flow.

Let America go,
Let it struggle with its fate, but
Let it flow.

A quiet voice is beneficent,
No bombs, no assassinations,
Let it flow.

Let it flow, Rifkin,
Let it flow,
Your piano, noble, American.

Let it flow, Joplin,
From your grave,
Give me your song
Day long.

STOPPING A KALEIDOSCOPE

The world is kaleidoscopic, ever changing,
Pieces falling into place momentarily
To give and fix a world of lights and colors.

Each stop is a strange new situation,
Each decade has its colors and its lights,
Each stop shows the world newly formed.

Abstract yet real, visual excitement,
The kaleidoscope shows, fixes truth without error,
So it seems, fresh views to the young, to the old.

The imagination to toy with this toy
Thinks no stop or view stays permanent,
Has to believe permanence of impermanence.

Eye and hand are too restless not to move it,
A slight or large movement resets bright scenes,
As time has been remaking the world to our view.

Decades shift and fall into place again,
Energies of the world are stable unstable,
Our senses camber the kaleidoscope.

I stop it, and there I see the world of the present,
Manifold reality fixed in a moment of time.
I turn it. It will be the twenty-first century.

Sameness ever the same, ever changing,
Growth and stasis in a hale bemusement,
We know what we see, see what we know.

No ebb, no flow not subject to change,
No changelessness in our consciousness
Remaining, nothing but what Heraclitus told.

NOSTALGIA FOR EDITH SITWELL

I wish it were the nineteen-fifties
And I was going to see Dame Edith Sitwell
And Sir Osbert in a restaurant in Boston,
And my friends would be there, and they

Would be sharp and handsome, friendly,
And we would think that the world was great,
And they would think they had a handle on it,
And I would think everything in the world is important

And Edith and Osbert were splendid and imperious
And somehow Death had not taken over the world
So we could be full of grace, with a sense of grandeur,
And O what deep realizations we had of life!

And after the shrimps and the wine and the haleness
We walked onto the streets of present Boston.
Osbert bowed to Dr. Parkinson, and bumped a tree.
He could not stop, so we had to stop him.

We lived when words were like roses on bayonets
With long-nosed Edith, and long-boned Osbert,
Alive to intelligence and to poetry,
Over their shoulders shadows of the Plantagenets.

GNATS

A society of gnats
Hangs on a beam of light
Near the ground, toward evening.

Then they rise up
And hang in the air,
Animatedly bunched.

What is their meaning?
I cannot guess their meaning
They are so ephemeral.

Nature makes them come and go,
As it does us. We
Amass our own society,

And I cannot guess our meaning,
Although I have tried for fifty years,
Twisting and turning.

SAGACITY

What is the use of being sagacious
If wisdom cannot conquer time?
We are sitting at tea in 2500 A.D.
I said love was good
Five hundred years ago.

My times vanished quickly in a mist
But here we are enjoying tea.
We are sitting at tea in 2500 A.D.
Love is still our principle,
Quite dubious, quite good.

If it were not for love we would
Not be begotten but forgotten.
The world is no better and no worse.
You thought of Dante as ancient,
We think of you as prescient.

A SNOWFALL

As the snow falls I brush it away
With a delicate broom so as not to use a shovel.
Every hour I go out to the long walk,
Conquer the new swirls and pile as if persistence
Were a virtue to keep up with nature.
If I did nothing I would be snowed in.
Some slumberous thinkers think this the best, January.
Let three feet fall, stay indoors, go to sleep,
Luxuriate in sleep like the groundhogs and gray squirrels.

There is something in me to test nature,
To disallow it the archaic predominance,
And if the skies blanket us entirely
With a silence so soft as to be wholly winsome,
(This beguilement of something beyond the human)
I have enough in me to give affront
And take my thin broom against the thick snowflakes
As a schoolmaster who would tell the children
What to do when they are getting sleepy and lazy.

I now make my predicament equal to nature's.
I have the power, although it is timed and limited,
To assert my order against the order of nature.
The snowplows begin to take away the snow,
Flashing big lights in the middle of the night.
They, corporate, have the same idea that I have,
Individualist, not to let nature better us,
But to take this softness and this plenitude
As aesthetic, and control it as it falls.

THE BONES OF COLERIDGE

O high hilarity,
O unexpected news,
O English gloom,
Almost Shakespearean,

Certainly Coleridgean,
I am told
By a British lady
Of authority

That the bones
Of Samuel Taylor Coleridge
Could be poked
By boys

Until sixteen years ago.
O high Highgate!
The Highgate folk
Wanted to keep

Him out of Westminster,
Out of local love,
But the churchyard
Grave fell

Into such disrepair
That boys with sticks
Could pester
The bones of Coleridge,

That this was a trick
Boys loved,
To poke a stick
And wiggle the bones of Coleridge,

I say
What a joke,
What youthful expertise
To prize a metaphysician.

A LOON CALL

Rowing between Pond and Western Islands
As the tide was coming in
Creating, for so long, two barred islands,
At the end of August, fall nip in the air,
I sensed something beyond me,
Everywhere I felt it in my flesh
As I beheld the sea and sky, the day,
The wordless immanence of the eternal,
And as I was rowing backward
To see directly where I was going,
Harmonious in the freedom of the oars,
A solitary loon cry locked the waters.

Barbaric, indivisible, replete with rack,
Somewhere off where seals were on half-tide rocks,
A loon's cry from beyond the human
Shook my sense to wordlessness.

Perfect cry, ununderstandable essence
Of sound from aeons ago, a shriek,
Strange, palpable, ebullient, wavering,
A cry that I cannot understand,
Praise to the cry that I cannot understand.

THE PLAY

As a gigantic harp in Vermont
Played in the hills by the air

Lifts and thrills the spirit to see
There in green rondure, a splendid thing,

Waiting for air fingers of sufficient force
For so large an undertaking,

So the spirit makes a music too, of me.

And I am played upon by airy thought
And make an air for you, an air for me,

My shape standing up in the hills
Tall and strange from millions of shoulders,

Man against the sky, man become a harp

Giving off melodious airs in summer time
And venting harmonious tones in winter

Received from heaven and given back

As if man could imagine a harp in the hills
To pluck out of him the delight of his spirit.

371

LEARNING FROM NATURE

While I was
Sitting on the porch, involved in air,
A small bird
Whisked across low,
Four inches from the floor,
Struck a glass door,
A rectangular pane,
With his bill head-on.

He stood dazed
While I looked on amazed,
The silence ponderable.
I wondered
How hurt he was,
Startled
By aerial reality
Piercing contemplation.

Plato was present,
Sophocles, Shakespeare,
Boehme was looking on,
And so was Blake,
Perhaps Dostoevski in a fit
And my friend Angelus Silesius
In the air.
They were interrupted.

The bird suddenly flew
Off into the darkening afternoon.
He did not say how he was.
He was stopped, and he went on.
It taught me acceptance
Of irrationality,
For if he or we could see better
We would know, but we have to go on.

THERE IS AN EVIL IN THE AIR

There is an evil in the air
Like shapes of powerful phantasies
Will wheel, will snap the trees;
It is the future, terrible, everywhere.

There by the cicada-sounding sycamore;
By the mill, off the hill
Comes the power to tear and kill
Into the heart of mankind more and more.

O dreadful evil, dreadful absolute
And furious thrashing words in the heart,
From the ramparts of art triumphant, I'll
The pure cold realms of death salute.

SOMEWHERE ELSE

Passion, too intemperate,
Goes off the track.
Reason, too reasonable,
Is tacky.

Where do we go from here?
Reason, too much, extreme
Passion, too much;
Lurch toward the Golden Mean.

Lurch toward the Golden Mean?
One should
Approach it gingerly,
As if an Absolute Good.

How could it be,
With lust for killing,
An Absolute Good
In disasters of our willing?

373

If saint and sinner are one,
The human condition is
Balanced. On one side contrition,
On one side derision.

I bow out.
If anybody has seen the light
He will bow out.
The light is beyond criticism.

The light is beyond distinction,
It is beyond division,
It is beyond question,
It is beyond precision.

THE IDEAL AND THE REAL

The word that burned into a page
Destroyed time, but the page is still there,
   Blank.

It was so immediate it flew away,
But reality was reality when I was young,
   Singing.

When I was young I destroyed time
While lovingly I was discovering it.
   Flashing.

I knew the absolute of the immediate
When I looked at a lily in the thicket.
   Heaven.

I knew nothing of heaven but of earth,
Earth was heaven at eye and fingertips.
   Unforgettable.

How could I know that I was obliterating time
Only that time should smudge the vision?
    Unsuspected.

The word I wrote on a page was absolute.
Everything was poetry, everything was real.
    Sensation!

The word burned into a page
That withstood the passion of my imagination,
    Another reality.

There was no death in the incredible garden,
Now age brings time on an open page,
    Inditing.

The timeless gives over to the labors of time,
Man's glimpse of perfection meets imperfection.
    Kingdom come.

TO THE MAD POETS

Happy are the mad for they are able
To see
That stones are light
That clouds are heavy
That women are men
And men are women.

Happy are these madmen
To see
That they were misbegotten
That they are misshapen
That they are free
Where others are chained.

Happy is their ferocity
Bought with painful recognition
In august cognition,
In terrible degree.
It is a happiness
Of spontaneity.

Happy are these times
To see
These madmen produced by them.
In guts and rubble of events
As from Islamic tents
They rush forth in their Fantasia.

Horsed under the maddest moon
Of ancient Africa
Rushing beyond the law
They shoot off their guns
In an ecstasy of Fantasia,
Revolving every dilemma.

LORCA

Soon it must come, the great charge,
The caparisonning of belief
And I will step to fight the bull
 A las cinco de la tarde
At five in the afternoon.

I will pass by his horn,
I will be brave and discreet,
The bull of times comes on
 A las cinco de la tarde
At five in the afternoon.

And I will daunt the beast
Near, near to my breast
In the heat of stealthy movement
    A las cinco de la tarde
At five in the afternoon.

And I will taste his breath
In the effort of the real
Whether of life or death
    A las cinco de la tarde
At five in the afternoon.

I will to meet my fate
In the quick play of the steel
    A las cinco de la tarde
In the strength of lust
At five in the afternoon.

A LINE OF VERSE OF YEATS

World-famous golden-thighed Pythagoras.
You can take apart the syllables but miss the thigh,
World-famous golden-thighed Pythagoras.
You magically incline into the line of music,
I say this line over, sometimes once a decade,
Or at any odd time of any unspecial year,
The marvel of it, the image without a totality,
No picture of the whole man, what he was doing,
No intellectual picture of a leader, or of a nation,
No rushing to a dictionary of Greek philosophy,
A transfixing experience, sitting in the dark in bed,
Or beside the bed at dawn in a dawn litany,
The magical, lulling line become a lullaby,
I heard it fifty years ago, I hear it now,
World-famous golden-thighed Pythagoras.

EDGAR LEE MASTERS

I am old, stony-faced, and hard,
But I had fire when I was young.

I began telling it like it was
Shortly after it was like it was.

It has never been so again
And I have never been so either.

I felt the real itch of the people
From swinging on their grave stones.

Now I have been made into a stamp.
I always knew I would take a licking.

THE PLACE

I

Eventually one finds
There is no environment
Patent for the poetic.

Any place will do.
Alas! One thought of a gold
Hullabaloo, a place of glass

Refinement with subtleties
Crossing the transparency
As lively as mind's images.

One thought of a vast portico
With appropriate, energized
Gods and beings, rich purposes.

378

Alas! Any place will do.
There is no poetical place,
America continues its practices.

Final toughness of the word,
The word bawling imperfections,
Its paradox to be heard.

II

There used to be
The violent struggle
For place, the right

Place poetic in countries
Or cities or underground,
The right place

Was thought emergent
And to harbor you,
Hello! Poetry Place.

The subconscious was
Nearest, perhaps dearest,
Anyway sheerest

But always fleering off.
Ways you went! Allurement
In echoic happiness.

There was no place for poetry.
Entrenched, my flesh is
Poetry's environment.

OLD DICHOTOMY: CHOOSING SIDES

Why don't you like the wild cry of the madman.
Who does not know what makes him cry as he does?

Because Aristotle said the world was measurable,
Took leaves off every tree, and measured them.

He began the scientific method. But the wild man
Was perhaps older, subjective, would scoff at the objec-
    tive.

We have to choose between the wild in us, and the sober,
The intensity of genius may be the best,

In it we recognize some true likeness,
Strugglers for change, visionaries of a bright future,

Keenest sensibilities eager for anything new,
Knowers of the first source of universal acclaim,

Shelley, Blake, and Lawrence knew of this essence
Of mighty realities thrown off as felt abstractions,

And why does the world have to be so slow and practical
As not to live for high vision instead of scientific clay?

And why was Plato deeper by far than Aristotle,
The far light he saw of absolutes, the Eternal Types,

Allied him with creativity at the heat of creation,
Even the new physics represented a principle of indeter-
    minacy.

And why do we cherish the mad poets rather than the
    sane
If we do not feel the truth of the immeasurable?

The truth of the incalculable, wild imagination
Where anything goes, and nothing is held back,

The genius of total illimitable, essential man
Who knows the joy of what it is to be free.

TRANSFORMATION

The lilac speaks without a voice: pure blue.
Weather has not killed it,
Nor storms riven it,
The sky has descended and kissed it into you.

You are the lilac hanging in the air.
You are blue speechless beautiful being, all fair
And graceful in escapelessness,
All lilac, blue-found beyond evasiveness.

OPPOSITION

Wildness of nature is in Florida,
Where senses open to the enveloping heat,
Where no Puritan lives, persons respond to excess,
The lust of the idea of Paradise.
Where Ichetucknee springs from limestone depths
Changeless through clear centuries,
As we float tubing down the enchanting stream.

Tameness of nature holds back the North,
Desolations of the past eroding the present,
The long, heavy shadow of the Puritans
Teaches the severe. They thought life could be better,
Prayed to God not to do anything wrong,
Held back their passion, aimed to kill,
Burned as witches free life-loving girls.

THE SWINGING BRIDGE

When an hour is harmonious
    Destruction lurks
    Inside the ear.

When you think of good life
    Inside your body
    A wild cell.

When in strength of love
    Imagine the weakness
    To come on you.

What of the thousand years?
    You cannot tout
    The meaning.

Lover to lover, eye to eye,
    Destruction lives, also,
    Inside the eye.

Walk across a swinging bridge,
    Make love in the afternoon.
    That's it. And fit.

MISTAKEN IDENTITY

The man who ate mushrooms
And tried to crucify himself
Cutting a stake, piercing hands,
Then feet, then hanging on vines,
Then walking home and stabbing himself
With a two-inch knife didn't make it.

382

If he got higher he had sunk lower
But instead of giving was only getting
When in confusion he presented
Himself to himself as J. Christ
It was pathetic. So mushroom busted
The bastard blamed the world on his ills.

If he had given himself to others
Instead of giving himself to himself,
Blasted on the heath of misadventures,
He might have reset the delicate
Balance of nature unbalanced by tricky
Gestures, he might have vomited mushrooms,

But, prized by the spirit of overkill,
It almost got him over underkill
Whereupon he had proved that hell
Is the joke of time, and been no man.
He is walking around sewed up,
A man beginning all over again.

ICHETUCKNEE

It is the continuous welling up from the earth
We must remember. Dawn comes, and the waters
Spring fresh, clear, vital from the earth.
Night comes, they well unabated from the dark.

Strange, is it not, that the temperature
Is always the same. The clarity is without change.
As the water blooms upward to become a petaled river
Each grain of sand below is visible as in air.

Over the oval, the mouth, the maw, the source,
We cannot see down into the cavernous mystery
Into primitive limestone releasing the clear water.
We are impelled outward from the warm, strong center.

Our bodies delight in the flow of original life.

Freely in the stream of exhilarating non history
We can walk, swim, float in the clearest shallows.
Upon us the welling up of source,
Around us the gift of the river, the way we must go.

Our bodies delight in the flow of original life.

HARVARD STADIUM

Sitting at the top of the stadium
The poet who was beyond this age, felt
Superior to the football players down below
Struggling back and forth on the turf.

He felt the elan of the senses, young muscles
Exhibiting prowess of the field of youth.
They were captive in their presences,
Held in bondage by youth's inferiority.

So felt the sage, his fame secure in age
While these youth could not foresee the future.
Some would win, some lose, in worldly struggles.
He was indifferent to victory or defeat.

Life was a struggle. He was known in the world
As one who had won the prize of worldly acclaim.
He sat above the battle like a god,
With lordly eye viewing the happenings.

No matter, now, who won or lost,
His musings unnoticed by the spectators.
He was a quiet victor at the spectacle,
The reddest judge of the rueful scene.

384

So God himself might sit apart, and stare
At the antics of the animal man,
Seen from above millennia of wars and non-war,
Indifferent to the players down below.

JOHN FINLEY

I

While three hundred Americans died in flight
From a broken nut, in cracked materialism,
Here on the stage a man appeared so fine
That he became the essence of immateriality.

He became iridescent in limitless belief,
His entire life became his luminous speech,
His face beamed that of the gods he loved,
His voice almost supernatural in its reach,

Yet human, the mingling of reality and unreality,
The vision, the passionate control of the ancient Greeks.
He became, in his astonished, firing mind

Plato and Socrates, he seized Sophocles,
There shone Odysseus the sufferer, Penelope alone,
And claimed the great Achilles for his own.

II

One man to speak beyond three hundred
Who died of a broken nut in a fiery crash
Of materialism, matter splattered on the earth,
One man, incandescent, to preserve our heritage

Of the greatness of the human past
When gods and men lived closely together,
When death was not meaningless, and harmony
Prevailed as love and wonder in all weather.

One man so rational and so keen
That he was transformed before our transfixed eyes
And became the gods that we had seen
When we were young to wildness, to wild cries.

He was a man radiant as history, as we
Stood to the glory of an ancient story.

EMERSON'S CONCORD *

Dear Ralph, to your Great Stone Face,
I would not have called you Ralph in your time,
The intimacy draws me closer to you,
You were here in Concord, your young wife had died,

Or was it the other Concord, this excites me
That we have two Concords. I do not know which one
Where your spirit hovers over the late fall landscape
And if it were the other one, the distance is not great.
Dualism as opposites has kept me from unison,
Unitarian simplicity ignores the grapes of wrath.

In my operant dualism I do not have to be Emersonian
Exclusively. I can also be Whitmanian,
And to break the mold in favor of no limits to poetry,
I can be Trinitarian, call Emily Dickinson Mary.

Dear Ralph, if you were an old poet at this banquet,
What would you make of our upheaved century?
You ventured as far west as Chicago, then daring to do,
To back American business, believe in our expansion,

_____

*This poem was first read at a dinner in Concord, N.H.,
honoring the author's appointment as Poet Laureate of New
Hampshire in 1979.

But what today would you say of World War II, Korea,
Vietnam, bombing Cambodia, fascist Watergate,
Our fear of law, our fear of impeachment,
What would you think of millennial radioactivity?

You had the word as in your time, Ralph, complexity
Of language made simple for all to hear, you teach
To sense the supernatural in the natural,
Wit remarking that Transcendentalism is a little beyond.

POETRY AND GAMES

I

The lightweight poets are happier,
They wave a silken handkerchief,
A necklace,
Toss up baubles here and there.

However, the heavyweight poets
Kick you in the groin
Disobeying the rules,
Wrestle you to the floor

Like the Golden Greek.
When they have got you down
They kick you in the groin.

They jump up and down on you.
Didn't you know
You should have been savaged before?
They say

There is no tragedy in tragedy
No fun
But smashing idols,
Truth is the work of fools.

II

However, the middleweight poets
Would like to smash you
But do not have the weight,
The gall,

They play it both ways,
Tinsel here, cynicism there,
It is all a game,
A world to make,

A laugh, a tent stake.
They play it straight,
A feint, a biff, no Frank Gotch.

They're skating out
Northeast
Southwest
West, east, south, north

Somebody is hurt, puck in the face,
He's in the box, no, he's out,
They're skating out, blue line
Timing, win, lose.

THE TRUTH

The only hope is to catch the moment as it flies.
We lift our spirits to this hope and salute the wind.
The wind is the reservoir of being. All go
Into the play of time on our flesh and bones.

We think to hold our loves and hopes intact
Yet when we love, love flies away from us.
When we feel that we have harmony and control
Cancer rips the life out of our lungs, or bowels.

Grace and charm, finitely flexible,
Make the old Viennese dance with joy as they
Sway to the music, revealing rhythms that
Capture spirit in delicate order and ardor.

Men and women know the fear of death,
When all shall be taken away from us, yet the
Inexpressible approached in poetry,
Careen of man in keens of music.

The wind of the spirit blows over mankind.
It is the vast gesture of the universe.
To have felt the pulse of deepest being
Is to catch eternity in the moment, life in death.

And the young shall try it again, the old
Regret they cannot express the inexpressible.
Fervor, effort, the dance, wild spirit
Flow and stay in the mind, and the soul

Shall worry the flesh to come up to it
And the flesh in ecstasy will deny the soul,
The flesh afraid of so great a prospect
The flesh knows it will die torn apart.

As the music sways us in strife and charm
The only hope is to catch the moment as it flies.
It flies from us. We cannot know
The ultimate purpose of either life or death.

Pick up the guitar, swing your girl, love the real,
It is all illusion that you think you know.
Necessity will take and shake you by the backbone,
The end for the living is a place too slow.

What hope, then? Catch your spirit out of the air,
Rejoice in resiliency of being, and being free
Sing the song of the song of splendors, sing,
See the light, sing, stop, sing, see the light.

NEWS

The news today is not bad.
It is that the sun has arisen.
This is the greatest news of the world.
Our newspapers give much bad news.
The earth is still riding around the sun,
We are earth-riders through galaxies.

The news today is not bad again.
The race of mankind lives today.
Men-women have not blown themselves
Off the earth, or blown the planet apart.
There are men-women of all shapes, temperaments
Who are living this life from life to death.

The news is that the moon still exists.
It is an instance of loving importance
Because brash man has now brushed her surface.
The moon has ceased to be a poetical virgin.
Will she bear good poems present or future?
Poets have yet to write a seminal moon poem.

The news today is good as to God.
Although some think that she/he does not exist
It is hard to beat the argument of Pascal.
He saw a watch in the mud; he deduced
That it had been made by somebody.
He looked at the world, and thought the same.

After God, we can only go to the devil.
The news is that the devil is alive and well.
The devil gets himself most in the news
Because we love him, and he loves us.
If I could say that order outlives chaos
It is because the devil reminds us of hell.

THE FLAG

The flag was a fabric to wave in victory,
Hold up in defeat.
It was the symbol of the tribe,
Tribe nationalism.

When World War II was raging
Each side held up is particular flag.
As millions were dying
Which side was God on?

We held ours at Iwo Jima,
A picture of belief.
Lee held his up at Bull Run.
All wars are man's defeat.

SEA STORM

Evening at the calm,
That's the best of all.
The seas quiet enough to think.
Not to have to combat them.
They are so much stronger than
Man they could kill him.
He survives, and triumphs, for a time,
By chance and wit. Wit to foresee
The fall of the barometer,
The danger of old charts,

Too many aboard,
Lack of ship-to-shore CB,
Lack of a young mariner
In case old ones
Have a heart attack,
The sea
Is not interested
In the pleasures of summer folk.

Evening at the calm,
If it would only stay still
Like this,
The full tide coming in
With a slight breeze,
The tide extraordinarily full
Under the full moon in July,
The moment of stasis
I praise,

I could tell you tales of the sea
And will tell you one.
At four o'clock
On an August afternoon
With a heavy southwester blowing,
Timmy Rhodes appeared,
A calm man on the coast of Maine,
Owner of Beach Island down Penobscot Bay,
And said,
Would you take me out in *Reve,*
My son may be lost in the storm.
He left Beach Island in the morning
In a slight dinghy with an outboard,
Has not been seen since.
He may be washed up on an island
Between Beach and the mainland,
Will you go search for him?

We knew there was not much light left.
We said sure, let's go. Timmy and I, and
Jackson Brodsky, six feet six, and our dog,
Boarded the cruiser *Reve* in heavy weather,
I with misgivings, but hopeful to find Timmy's son,
And took off toward Spectacle Island.
We searched the shore, found no body or boat.
We headed for Fiddle Head and Hog Island,
The seas were high but still island-repulsed.
We searched Hog and Fiddle Head.
When we got beyond Hog heading for Beach Island
The full weight of the storm bore on us,
From the West over the Camden Hills,
To left and right a hundred miles of ocean
Came at us in a red ominous light of sun
And man-chewing fury.
We took the seas on our starboard quarter.
Half across
Tons of water crashed into the after cockpit,
Fortunately self-bailing,
Whereupon Timmy said,
With the calm of an old New Englander,
"I wouldn't let that happen again."
I had heard the phrase about
Shivering the timbers. *Reve's* timbers shivered,
She rose up and shook off the heavy waters
Piling over her and into her,
One stood fast at the wheel, all one could do.
I plied the cruiser like a sailing vessel,

Turning her into the highest waves,
Searching out any point of vantage,
We had faith in our vessel,
But knew to expect any eventuality.
The red sea in all its might and selfhood,
A deadly sight of malevolent oncome,

393

Total affront, how could we survive this,
Yet we kept on through the forcing waves
At low speed, and came to Beach Island
Where the storm was so strong,
The waves so high, we could not
Pick up a mooring, had to
Head her into the wind,
Keep the engine going, keep her
Head on, direct against danger of sidestroke,
And how long could we do this?

Timmy, fearful for his son,
Wanted to get off,
Go by dinghy to Lisa Jane,
Insisted he would not give up
But search the islands for his son.
We didn't think he could get off,
How he got off I'll never know
But he did, and rowed to Lisa Jane,
Took off in lording seas to find his son.
We might never see him again.

Now the light was leaving, we had
Only about an hour, and had to decide
Whether to seek the leeward of Beach
And anchor for the night, in the tearing,
Tear-forcing storm,
Which we knew not whether it would
Get worse or abate,
So we left perilous stasis
Of heading into the wind
And a hundred miles of red, hard-evil seas,
And got around to the lee.
It all seemed desperate.

We finally decided
That since we had crossed from Hog to Beach
And taken tons of water,
But survived,
We had better chance it again,
Get home by nightfall in half an hour
Rather than chance
A slippery anchor
And the unknown terrors of the night.

So we headed *Reve* into the storm,
Now with the wind and waves on our port,
And tried her through the waves
Evading the big ones turning into them.
It was a poker passage but we made it,
Some lee help from Fiddle and Spectacle,
And brought *Reve* to mooring at Undercliff,
Just before dark.

The father who might have lost his son at sea,
Timmy, appeared later at Undercliff by car
From Buck's Harbor, with his tall son, who,
His motor conked out, took off his shirt and
Made a sail in the dinghy and in the high wind
And seas, a young man sure of himself, fearless,
Sailed into Buck's Harbor miles away,
Hardy and able, evading death, even happy,
And Timmy had scoured the faceless seas,
Not found his son or boat,
And directed Lisa Jane into Buck's Harbor,
Where was his boy
Luckily out of the predation of the ghastly waters.

We had a round of drinks
To fortune,
That happened to turn out good that day.

395

KEY WEST

Is far out, umbilically extravagant.
Pelicans come in behind the shrimpers,
Tossed cutoff fish heads from the day's catch.
Dogs bark at night, cats high-pitched.
The sun is the greatest thing about Key West,
A savage source. Dangerous to the thin-skinned,
Blonds fight off its cancerous attraction
By moving to shades whose winds may cool.

Audubon's done-over house has a shady garden
With cool white iron chairs. Hemingway's house
Is more elegant than you would have thought of him,
The cat population has declined from fifty to forty-one,
Strewn placidly about, some authentic six-toed descen-
    dants.
Tennessee Williams' house is entirely closed in,
A tropical hideaway surrounded by walls.
The graveyard looks like Italy not America,
Houses run from quaint to spectacular, to old.

GREAT PRINCIPLES ARE THROWN
DOWN BY TIME★

You stand thinking of great principles
But they are thrown down by time.
You think your intellect holds them
But your intellect is altered by time.
Time changes, the eyes change, as fate
And the great principles go on in time.

★This poem was read by the author to Vice President Walter
F. Mondale at the "Spirit of the City" award dinner of the
Cathedral of St. John the Divine, in New York, in De-
cember 1977.

What if Aristotle, if Plato were here?
What if they were at the Academy in New York?
Can we imagine it? Are any of our men
Able to withstand thousands of years
And enounce principles that are the great principles?
We freshen ourselves in throes of difficulty.

Yet the great principles of the live Greeks
Are thrown down by time. I heard
A commissar in Washington in 1960, when I
Got too friendly with Yevtushenko and Vosnesensky,
Discredit the entire classical Greek establishment,
Say the tragic flaw is a defunct idea,

There could be no tragedy in his country
Because all were equal, none could ball
From a high state to a low state because of a flaw
In character; like bees or ants, if numbers
Were killed numberless numbers would follow
To keep the hive or hill alive with future.

Individualism inheres in Western freedom,
Dynamism thinks it knows which way to go,
Blake, Keats and Hopkins shine alive,
Democracy's skin burnishes to one color,
If we want to know what is the matter with America
Now, we can warm our guts in the blood of Whitman.

We can feel the vitality of our striving
Here and now, in the heart-city of our land,
Can sense the visionary grandeur of our founders
In the erected freedom of our high-sprung spirits,
And when we suffer and fall, hail to America,
We have the strength to throw the evil-doers out.

Great principles are replaced by time
While the eyes see man as teeming, an upward
Animal, begetting in love,hope, and belief.
If he joins the brontosaurus, the Incas, the Aztecs
So be it, and so be it love incomplete.
With poetry he was replete.

THE LONG SWING

I miss the long swing
Under the great tree,
Which reminded me
Of my youth.

Back and forth,
Forth and back
I went in the long swing
In my age,

Upheld by a giant branch,
Like the son of a father.
The swing was metrynomic,
The present whole and all.

Grace and charm,
The slow, long glides,
Blood's harmony,
Sense of long love.

HOW IT IS

Then the eighty-year-old lady with a sparkle,
A Cambridge lady, hearing of the latest
Suicide, said to her friend, turning off
TV for tea, "Well, my dear, doesn't it seem
A little like going where you haven't been invited?"

GOING TO MAINE

Going to Maine is a state of mind,
Like everything else.
You may have been on Guemas Island,
In the State of Washington,

Viewing the Cascades wide over water,
Watching an eagle soar,
Impressed with the quantity of water,
And eaten bear steak with the McCrackens,

But when you return to ancient New England
The first question asked on Main Street,
With breathless expectation, is,
Are you going to Maine?

Are you going to Maine, oh,
Are you going to Maine?
And I say, yes, we are going to Maine,
And they say, When?

They want an ultimate answer
To an ultimate question.
Pestiferously human,
As if to infestate inner skin,

They question, almost with a triumph,
When are you going to Maine?
As if you were going to Heaven
And they would see you there!

And you say, yes we are going,
Harsh to be indefinite,
Yes, we are going, we are going,
Yes, we are going to Maine.

SPITE FENCE

After years of bickerings

Family one
Put up a spite fence
Against family two.

Cheek by cheek
They couldn't stand it.
The Maine village

Looked so peaceful.
We drove through yearly,
We didn't know.

Now if you drive through
You see the split wood,
Thin and shrill.

But who's who?
Who made it,
One side or the other?

Bad neighbors make good fencers.

COMMAS IN WINTERTIME

The cold holds everything in abeyance,
The effort to find out the mystery of life,
The struggle to perceive what is perception,
The longing to pierce the truth of St. Theresa,
The belief to discover what should be belief;
The radical daring to dare at all,
The grandeur of messages from another world

While the intellect says there is no other world,
The pain of realizing another death
When we are headlong going to our own death,
Prospects of desire, thought to be illimitable,
Caught short by the failing of everybody,
Shoots of remembrance, long and oblique,
Of incredible joys lost beyond memory,
What it was like to be truly enraptured,
How the senses captured what was remarkable,
How everybody else would die, but not I,
The leap of imagination in the dark,
The grasp of self, else lost to others,
Hopeless belief in one's infallibility,
Darling ego, surge of certain selfhood,
How could the world go on without me,
One's surge love never to be recognized,
The inner secret of inscrutable events,
Soul of the world, ever to be secret,
Reality, hard and final, final, hard,
Breathe shallow, it is twenty degrees below zero,
When six I saw Halley's comet over the back fence,

FANTASY OF A SMALL IDEA

I have a fantasy that a small idea
Is as good as a large idea, may be better.
Einstein had a large idea, but he begot
Possibly the blowing up of the human race,
So it could not be called such a good idea.
But maybe there is the little idea of love,
So little in our time as to be debased
From what the ancients thought of it as grand,
And as Freud belittled it by dissecting it, .
And who with Satanic Hitlers and Stalins,
Struck great ideas of the world down

401

Which were announced by the ancient Chinese,
By fifth-century Greeks, and by Jesus Christ,
As well as by Mohammed and Buddha,
Maybe it is time before atomic holocaust
To fantasize that any small act of love,
Say any goodwill eye-flash to a passer-by
Is just possible a great gain to humanity,
That to love anybody is a triumph of instinct
And if there are enough small acts of love to save us

We might outwit perhaps dream-bombing scientists,
Even take care of our planet without stabbing and killing.

A DREAM

A loon's cry is a chortle from another world.
Gluts of silver, the dawn conclamant,
The ruffian band appears at our house.
Cinematographic, they move in,
Each face set in a rigid throat,
Their unity impressive and ominous.
The owners hover in an upper room.
Two knock, say they have come,
They need not say it, to destroy our house.
They break plates all over the place.
Their youthfulness and zest is mastery,
Without qualitative argument.

We do not have to argue either,
But touch relatives with quick glances.

In a room chock with swarming braggarts,
One is perhaps startled as two abreast
A column forces in with long guns vertical.
Our sin is putting rouge on our faces.

They march with marionette absolutism,
When they get in they dissolve in the ground.
The leaders are arguing at the pedestal.

I am waking in a university city,
The halls crowded with brutal faces,
Hundreds force into the large lecture,
Dr. Faction is lecturing on culture.
He cannot be seen, cannot be heard
Behind the solid mass of twisting flesh.
In an anteroom women gossip and knit.
It is said he is being transferred to Harvard
To the greatest university on the continent.

The loons are savage and absolute,
Their cries annihilate the relative.
Our house is being destroyed, the crowd
Is dancing and mounting in a high glee.
We are ashamed of an old order
Of sanity for which it is useless to contend.

I have fled to a new adjacent city.
Two men are struggling with polar bears,
Each has his polar bear in his arms
As big as a dog: each is wrestling
And wrestling his away from the other.

One denounces the other, "You know
You stole my polar bear." He accuses
With righteous anger. The other fades out.
They dissolve in an intrusive symmetry.

By me
Is a jewelled reindeer bright and tangy,
The flesh of another world inviolate,
Attached to a sledge of violent colors,

It is a reindeer taxicab. I ask,
Shall I take the reindeer taxicab
Back to the consequential city?

The past of abandoned truth fades,
The new dawn appears.

TESTIMONY

I was going to make something of it
      But I lost the track.

The creative bright new world bright worldness
      Was there within a breath

Within a hair's breadth, a whisk of time,
      Some incredible minimal essence

I almost had the grasp of it entire,
      I leaned into fullness, meaning,

I had a glimpse of totality of experience,
      On the verge of an absolute,

Something sensed so fine as to be indefinable,
      Glimpse of godhead truth,

I was sure that it was going to be complete,
      As if my life were at stake,

A kind of state of being without resolution,
      Rich with immediacy too delicate,

It was a given sense of total selfhood
      When supreme vision was to appear,

404

The flash of some holy kind of instance,
    Some revelation imminent

When the elusive juice of electric spirit,
    The instant flash of knowing

Vanished from the mind with lightning speed,
    Left me at a lower rate,

In this lower state I lived alone and true
    To the vast corporeal thisness

Of thisness, each thing, each happenstance
    As solid as the world is long, and you.

Then I thought to invent inmost praise
    Of sundry matters everybody knows,

And feels of obdurate sameness, the common
    Touch that never effervesces,

Thought to praise life as it its, incomplete,
    Because praise of the highest vision,

Unattainable, glimpsed in a high moment
    Is altogether unattainable,

We are here in a ground of earth structures,
    Every day bread, workmanship,

We are beings who are held down to time,
    Amazed at a glimpse of immateriality,

We know we will exit soon and disappear,
    We had a sense of some otherwhere,

We had a sense of indefinable vastness,
        Beyond our powers to endure.

This is my testimony of that bright time
        You too may have felt sometime,

But why should this century deny me
        Capture of essence and light

When the Greeks and the Elizabethans knew them,
        Keats knew them incarnadined,

We are the materialists of the atom bombs,
        Fear seizes us in the joints,

We think a vision of immateriality
        Must have no meaning, none,

In our teeter and balance before annihilation,
        The end of us,

When it comes, when it comes, the blast,
        Destruction of the best and worst,

We wanted to look in the eye of God,
        We got six feet of radioactive sod.

THE KILLER

Rise up, poets, phalanx of the just,
Correct the stance of the erring man,
Be there in sunlight, take back his little shots,
Reverse the shattering camera of time.

There is the shadow of Oscar Williams,
Whom you left out of your anthology,
There is Gerard Manley Hopkins, who saw
A stippled trout, hurrahing for harvest,

Looking, saturated from the White Horse,
Dylan Thomas' world-view through the foam,
Yeats' knuckled hand move slowly, true,
Dowson will turn his back on you.

Stand to your last, and lock him out,
Poets, rise up, and stand in the light of day.

There goes magisterial Bridges, stately,
Long-lived, to suffer for a dying child.
They are closing ranks, are statuesque,
Thomas Hardy peers down to the Titanic,

Housman, no brute and blackguard, musing
Sardonic, himself a dying nonathlete,
A. E. fingers angels in his pocket,
James Stephens brings on a lady of the street.

They are forming for a picture of the times,
Eliot with the look of a Missourian,
Pound somehow radiating Idaho,
Frost quixotic, shifting on both feet,

Robinson Jeffers, claiming mountain, sea, and stars
Superior to the antlike crawl of man,
Hart Crane, enlarged on the Elizabethans,
Killed by ocean water, from Chagrin Falls.

There stands Edna Millay, called Vincent,
High on a hill in Camden, looking out,
Marianne Moore, pristine, in a tricorn hat,
D. H. Lawrence in a sandy-red beard, pointed,

Heavy, sun-faced Wallace Stevens, smiling,
Big-browed Muriel Rukeyser looking like the people,
Aiken battling it out with Freud,
Archibald MacLeish understood the Trojan horse.

407

Cummings troubled by clarity and evenness,
William Carlos Williams patting poets on the back,
Rexroth beating a drum for pacifism, eternity,
Ginsberg howling for mercy in the face of death,

Duncan, off in the hills quoting Milton,
Ferlinghetti, concerned about S. Francis of Assisi,
Corso closely, long, final on marriage,
Kees getting lost, a dropout, a cliff dropover,

They are walking in from all points now,
The poets to form themselves into a picture,
Some have come, many are left back home,
Far back in time, or in some other countries.

See, there, some of the fresh-strengthed young,
Lumbering Lowell, artful Wilbur, big-bear Roethke,
Behind them classic front-lobe Tate, vision-crested
    Warren,
Ransom the teacher, subtle of grammar,

Spender, who saw a high spiritual essence,
Auden, who would not flinch at the world's dirt,
MacNeice, who went for a skirt in a taxi,
Day Lewis, who wrote mysteries under another name.

They are gathering into a humane phalanx,
Each an individualist, given a gift and giving
The gift to the world of waiting takers,
Each is a discoverer, a knower, purveying truth.

        Until society teaches him your worth
        He will go on killing the President.

Luminosity of the aggregate poets,
If the principle of indeterminacy holds,
This poem could turn to composers or painters,
To sculptors, filmmakers, Chinese haiku inditers.

The idea is of ideal superabundance.
The world is so much greater than any man
Or woman's mind in it, greater than expectation,
That we do not wish nothingness too soon.

Think of the hard work done by Audubon,
Think of Stefansson eating only meat,
Think of Scott at the South Pole by horse,
If they had thought of dogs they would have survived.

Think of the wildness in the heart of man!
Mt. Everest does not care whether anybody climbs it,
What is a name to those eternal heights and snow.

Susan Butcher was first to climb Mt. McKinley
With dogs. My radiant niece, young woman bountiful,
Ran the Iditarod from Anchorage to Nome, a thousand
    miles,
Then dared, with a sixty-year-old, to see the Guinness
    book.

Her sister Kate, quiet, took to Greek and Latin,
Won a Greek prize from New York University.
Abundance! Is it better to take dogs up McKinley?
Or to understand subtleties of Greek literature?

What a joy to talk about persons not in the books.
Our poetry is studded with university acclaims and no-
    tices,
Poets are elitists of an unelite democracy,
Nobody will listen to them but those in the know.

Ideal superabundance, the state of grace of the nation.
While everybody reads the savage, low-lying press,
The ecstasy of primitive, inveterate consciousness
May be inflaming a young unknown poet to greatness.

Behind the news is as big as behind the eyes.
For behind the eyes, before they become blind,
And make us weep, but dry, no tears to weep,
Behind the news is the astonishment of reality.

The heart of the nation thrives with living thrust,
The heart of this nation is extravagant and kind.
We are a people who love and hate, hope, but kill,
We are the proud inheritors of the Western world.

To talk about an individual act of assassination
Is to talk about everything, to see our splendid heights,
And our depths of depravity as one and the same,
We were revolutionists first, we still kill.

I suppose murder is as good as kissing, a terrible
Thing to say, why does anybody murder anybody
Unless because of a self-love too great to bear?
Ideal love, too, is unbearable selfhood.

I would like to have my mind range free as a bird
And I would like to be able to put down this line
Before I forget it, because I have often noted
That reality is lost if there is not pen and paper to hand.

Infallible indices, I love the idea of infallible
Indices, as if one had all of life in hand,
As if an index could be expanded to perfection,
As if the dreams of Sir Walter Raleigh were not in vain.

As if the worm, the arrow, the poison had not
Contaminated us beyond redemption by poetry,
Whose source is, was, and will be ourselves,
Our vision does not want to be obliterated.

Perhaps obliterated somewhat, so that we see
Through a gauze not plainly but with evocation,
And we have to trust the mist of the eternal veil
Between knowing and the truth of unknowing,

How does all this sort with lavish reality?
Is reality so great a thing that we think
That reality is everything? The greatest things
Are full of unreality, as Christ on the Cross.

And if poetry could make you happy
It would have to contain all pain, all misery,
It would have to be deep in the substratum
Of your being that you would have—what?

What word can assess the human situation,
The lyric flight of the bluebird's joy,
The awful recognition of a cancer patient's death,
The changes of life, recognition, strengths of hope.

Think of our little cinder somewhere whirling
Beyond the consciousness of man in a death
No man can understand or supervise,
Nothingness the everything we can contemplate.

What was the hope of mankind, where
The territories of love, desire, human greatness
When language was silenced in a long darkness,
When God, inscrutable, is beyond the universe.

Here the song the poets sing,
It is the song of the ages, now.
They sing the moment of absolute good,
Piped by a piper piping.

They are to make the world anew.
It was true before, it is still true.
If winter is white, spring is green,
A spirit of being lithe and lean.

Go to the primal spring
Against killing anything.

Tick, zip, zip, tick
In the humid night air insects
Hit an electrical grid, emit
Zipping flashes big or little

As they zip off to the eternal
Flashing crispest sounds, ticks
Continuously intermittent,
Hit, flash, emitted tick

They encounter instantaneous fate
Hitting fatal electricity,
Tick, zip, zip, zip, tick.

If mankind can kill off the insects
Frisked in a flash, mankind is fated,
It is a matter of timing,
We do not make fireworks at night.

Nit, wit, tick, nit wit,
Flick, hit, flit, wit, nit.

A little spot of grease goes splat, life
Winks off. We are thrown against a grid,
Death, the noise of our death splutters,
Instant show is over but the show goes on.

An amorphous grouping, coming into consciousness
Is the standing picture of the unarmed poets,
Who, word lovers, lovers of the real, the good, the true,
Are armed with poetry to take life seriously,

Not kill another human being. Letters
Are their bullets fired into the brain of man
To sanctify and fortify, to enrich not kill,
They are a crowd of life, a phalanx of perception,

I thought of Baudelaire, Rimbaud, Mallarmé and kin
All in one line. I had to leave a lot of people out,
The picture had to be more or less of my time,
A few decades in a total, massed world-wealth.

A gathering of the spirits. What a throng from deep down
In the beginning of consciousness from childhood on,
These makers of life were makers of masterful meanings
Recognized by total populations of the bright.

They are summoned to stand amorphous but newly real,
Each beside the other, although they may not have known
Each other, all of them bound by wisdom and by love,
Their spirits materialized in marching words of evoca-
       tion,

Their messages loved by many, their lives
Whether harsh and bleak, or radiant or sunny
Studied by aspirers, the thoughtful, the knowing, those
Whose imagination mates with their imagination,

I cannot name those from fifty down, there are so many,
Nor can I see all of the forms of the mature and elders,
It is a pity but all limitations strengthen us
To perceive the vastness and greatness of what we do not
       know

413

Or own, but now I come to my austere belief,
The phalanx of these poets stands apart
At any man who aspires to kill the President,
They should aspire to read poetry instead.

Can madness be educated out of these youths,
In the future, these madmen, these killers,
Can our society tolerate their malign brains
When there is much in life they do not sense or feel.

It is improbable that they will see my picture
But we must strive to put an end to death
By gunshot, and earn deeper meanings by deeper know-
    ings,
The offended poets, standing in a row, shot too

By ignorance, fear, and violence of the killer
Cry out against the cold, barbaric bullet intrusion,
They stand here every one giving love and belief,
Hope, courage, strength, true mortality from poetry.

    Killer, you are a man. Listen to the poets,
    Or you will go on killing the President.

SAILING TO BUCK'S HARBOR

    The spectacle of the New York Yacht Club
    Coming into Buck's Harbor
    After diminution of the wind,
    Lateral sun rays held on expressive hulls,

    Ingathers yacht lovers in late summertime.

    Old-timers linger on the lawn
    Cocktail-handling, remembering

414

A same sight similarly suitable
Of tall ships coming in long ago

Before north sundown, now grandchildren are here,
Before the yacht club lecture,

"Sailing to Tahiti."

CEREMONIAL

If you don't go you don't go.
If you are there you are there.
You sit here. You are veritably here.
What is the truth of here or there?

If the present is here, is limited,
You think of there as a great show.
Yet since you are not there, but here,
Which is greater, a worldly, or unworldly time?

See, you have just won the Highest Award.
They stand up to escalate ovation,
But you are only a two-legged, forked creature
Soon to die and be nothing at all.

The ceremonials of success are ceremonials
Of excess because we have to believe in man,
And when he or she is given the Great Prize
We are aghast because of our loneliness.

In our fever we have put a blaze
Of approval on somebody we hold important,
As vain an exercise as we can imagine,
We do not have faith in our own death.

So, put down death and put up life
And hand a banner in the air
And if time and wind blow it to tatters
Fame ceremonial we think, matters.

THE SACRIFICE

Every poet is a sacrificial spirit,
Every song he sings is given
By special election in heaven
So that men may bear it.

Each toils, and throws his life away,
Gay as a boy tossing his cap up
For whether of tragic things and heavy
He lives in the senses' gayety.

Like some drunken bee plundering flowers,
Drunk with his gorgeous nature,
He gives a golden summer afternoon
Its fiction, intellectual and sensual.

THE LAMENT OF A NEW ENGLAND MOTHER

Where have I lost my way among money and horses?
My mind is like the keen edge of a blade.
I am a Bourbon of Vermont. My children are prickly in
    the wheat.
In the castle of torment I swing in the winds of chance.
Do I dare my Adversary to duty past delerium?
Do I mock the green ancestors in my recklessness?
Is there any recourse when death has taken my beloved?
That cancerous fiend has made my assertions vain.

416

The world is rocking that most stable was.
My estates panic in the trembling of my will.
I am the crossed and vexed one, the soiled evidence
Of universal malevolence, guilty to have been born,
To marry, to bear children, to affront Providence with
    spleen.
I am in search of a soul revolutionized.
When shall I see the pure stars of my childhood,
When shall I trust in the love of my pure husband?
When shall I unseat my selfishness, my false debate,
And live in the rich simplicity of the earth?
The world has visited me with viciousness
And all my life is humorless and viscid.
I cannot cope. I am the lost cornucopia
Of June. Yet I seethe with rebellion still,
Daunting society in the mazes of my perfidy.
Let me go, Fate, and bring me back to douceur.
The graveyard on the hill that holds the bones of my hus-
    band
Affrights me with the rancor of life. My lovers
Have all gone into the garden. My richness of fantasy
Plagues society; I am a checkerwork of secrets
Knocking together in a burden of black action.

IVES

Charles Ives, I have heard
The magic of your music,
Flowing day-long evanescent
And brilliant-scintillant

And I have whelmed in that
Many a day-dream puissant
Where your lithe-deft energies
Met mine in gayeties sent

417

Thanks to time this-ward
Fuming with deep pleasantries,
The bounding shifts and carols
Of the perfidious seas,

O Charles of the eccentricities,
American to the whimsy, blazoning
Our beliefs, fictive absolutist,
You engender profundities,

For, sitting to your portrait,
As the world goes winning and sailing,
I am a sage and lively trial,
Nature breeds and teems the while.

MYSTERY OF THE ABSTRACT

When the great nothingness was everything
We were nothing and had everything.
Think of the nothingness of our pre-being.

Think of the nothingness we are going to be.
Yet when a sperm hit an egg things changed.
It violated the principle of nothingness.

Things changed, time became, we began to grow and ever
    since have grown into questions
About the difference between nothingness and being.

Now in my eightieth year I sit by the shore
And admire ocean tides whether coming or going.
The Greeks knew Platonic truth at one time,

The Elizabethans knew another truth at another time,
We Westerners have learned to manage the world
But where in time and space is our planet going?

No reality but things said William Carlos
Williams. A tag. Make it abstract, ideal.
Soon we will be abstracted from life, be nothing,

Time, space, live, hope, death are overlays of life.
These masterful ideas make us human as we are,
No one escapes them, we have to believe in them.

When we become abstracted from life, are nothing,
We will be everything, everything man has felt and thought
Melded into eternal being abstract and ideal.

WHITE PINES, FELLED 1984

Not one not dandled a man high up in the air,
A man in his prime, a spider on a steel line
One hundred feet up on a hundred year old pine,
If he'd miss just right the whine of the band-saw
He'd cut off his limb, the high man a dancer
Against the trunk, master of the slight hand signal
To the landed man on the marvelous machine,

In a minute a bright round spray of sawdust,
Like a stream of yellow life-giving juice,
Signals the downfall of a hundred year old tree,
Even one hundred and fifty, gone all, gone in
A jiffy as she topples in direction prescribed by
A rope tensed to a tree down hill down a slope,
She snaps off just right and lands with a big bang.

The end of tree makes thud-rumble, moves the earth.
A line of magnificent white pines, one got out of line,
Snapped thirty feet from the top in a high wind,
Landed on my neighbor's garage, ruined his portico,
If it had snapped half down or more it would

Have creased the whole white house form attic to cellar,
Perhaps struck his eighties old mother moveless inside.

We knew what we did when delved or hewed,
Racked and crashed the growing green, rational with
A saving ritual so that not our house next door
Should be cleaved by a two-ton tree in a tricky wind,
Perhaps end us in our house, these things of beauty
Growing a foot a year become things of danger. Thirty
Years of beauty might produce life-crash, instant death.

When beauty is dangerous a man can dandle
Himself up high on a hundred foot white pine tree,
Able, without emotion, rational, the old tree at his
Lack of mercy and delicate materialistic apt know-how.
When beauty is dangerous top the tree, then bring
her down in resounding earth-wallop. Aftercomers
Serve cocktails from the beauty of the elegant stump.

A CLERIHEW FOR ALAN GAYLORD

We love to think of Alan Gaylord
Hopefully never gored by a sword
For all of us happily sent
To be the head of our department.

Who has not remembered Alan Gaylord
In shirtsleeves when he almost roared
At Psi U or another fraternity house,
Everybody almost in a beer souse.

It was this same hopeful Alan Gaylord
With whom all were so cheerfully in accord
That they relished his Wife of Bath's tale
The more because of the beer or ale.

420

We never think of middle age for Gaylord
But the middle ages are his hoard,
His high spirits are youthful,
The past he presents is truthful.

For department heads like Alan Gaylord
The clerihew was exactly tailored.
No part of the job was too hard for him.
Though the chance of winning was sometimes slim.

He was placed as clerihew is in Webster's hoard
Between the cleridae of administration boards
And the faculty clerisy often governed by whim,
Alas whose sight is sometimes dim.

With such a captain the ship was sailored
And came through the reefs by grace of the Lord,
It's good he had Beverly
On whom to lean heavily.

The sweat that he poured
Looking after his wards
Fought the predacious claridae whose checkered colors and
    luster
He met with hardly a fluster or bluster.

The love that has been stored
For our gallant Gaylord
Is clear and free
Despite the cleridae or clerisy.

THROWING YOURSELF AWAY

To throw yourself away
Is to throw yourself into everybody,

A part of embracing the world,
Not self-embrace, close embrace, wide embrace

As if everyone in the world were open to love,
Your love was unasked for, a gratuity, essential

To be thrown among mankind indiscriminately,
Like seeds of thought propagating the future.

You throw yourself beyond the world as well,
Not only among men and women, but among stars,

Galaxies, futures of space illimitable, time
Perceived so continuous as to be timeless,

You throw yourself away because of imagination,
Belong to earthworld, not any nation,

And long to think of a thousand years as a day,
It is a passion to say, that has no stay.

RIVER WATER MUSIC

With the sun half way down the tall pines
Heading for the earth of Vermont,

I sit in New Hampshire high over the Connecticut
Listening, eskered, to electronic music expand,

High decibel, confronting nature with man's extrava-
    gance.
Power burst of electronic music burst over the hills,

A kind of gigantic extravagance unknown to earlier ears
Which hear the rich slight wind whisperings of near sun-
    down.

Brisk, strong, air-penetrating reverberances
Fan out into the innocent air for miles around,

Man denting the idea of ideal silence
With power gestures waking the ears to new feeling

While all over heaven flaunt the claps of electronic aggres-
    sion
Saying the mind may be old but the sound is new.

THE WILD SWANS OF INVERANE

And we came near Corofin through a narrow gate
Down road where farmers were topping hay stacks
To Inverane on Inchiquen lake
To the house with the wind in the chimney,
The wind blowing all day, and days,
In the wildness of ancient history,
The wind hale and strong over the Burren,
And saw the island in the white-capped waters
And Moira put turf in the fireplace
That friendship glow the more, the slow fire
That beat in our hearts for love and history,
For the untamed nature of myths and sacrifice
Arise to the visions of reality,
And the great winds rattled the whole house,
The lake passionately breaking on the old shores,
And then to be startled by wild wonder,
Seven wild swans descending out of the heavens,
Out of myth, sacrifice, out of ancient ecstasy,
Settling with rough vigor on turbulent waters
In magical union of reality and dream,
Magical wild swans of Inverane.

SNOWY OWL

A poem taken by the New Yorker
About a snowy owl who came from Canada
From the northern tundra to New Hampshire
And lit on the cupola of an old barn,
Which was a century-old town landmark,
Settled there in snowy whiteness.

The word got around of an exotic visitor
From the Far North somewhat early in season.
Soon there were cars parked ten in a row
With citizens interested to see the apparition,
Most with the naked eye, some with binoculars,
A sight most astonishing in its impact imprint,
A bird, as it seemed, almost from outer space.

The snowy owl dominated the scene for days,
But in this barn, a century or more ago,
A man who had raped and murdered a girl,
One of our own in a ravine nearby,
Hid under the winter hay a long time,
Finally was quarried and, as it was, quartered
So strong were the town's feelings against him.

Many professors went to the public hanging in Concord.
Discord not concord brought this about,
Who knows what of preFreud caused this event?
Aeschylus, Sophocles, and Euripides spoke about it.
I gazed at the bird a long time whitely.
And report, finally, that the barn is razed.

GOING BACKWARD GOING FORWARD

(For Andrews Wanning)

Toccata's sails were up and set,
We were coming in from undefined ocean

To the beginning of large and small islands
Worked through before coming to far landfall.

Toccata, elegant and able to sea-play,
Lived on the waters with living joy, when

I noticed a marker to starboard going forward,
Impossible, we were going backward going forward.

Electrified at this astonishing revelation of the sea
We sat there peering expectantly toward the mainland

While we were slowly being carried backward out to sea,
Seeming wind outdone by currents and the tide

That kept Toccata shaped elegantly to the North
While, gently daft, Toccata slipped directly aft.

We were being sent beyond our wills backward South,
Slowly backward, graved on ocean, far out.

LISTING

I felt life coming on
Like a stealthy small zephyr,
As if something were imminent,
But I did not have a piece
Of paper.

425

Wildly bewildered, no paper to
Write on, life was coming on,
Inviting, enticing, incredibly fleet,
It was already flying away, no paper
To write on.

Such imaginary feats and excellences
Ravished my being that I could not say
What they were, or how extraordinary,
Because I was fumbling and could not find
Paper to write on.

Then I thought of my very self, and you
Who would read the paper that I would cite
And had a belief that this was tricky
Asking you to believe my mystery
Without poetry,

Wordless poetry flowing right on
Without anything to write on.
Could anybody be so eventual
As to live only in the sensual
Anon?

Or be so conspiratorial
With the unincorporated ethereal
As to raid the world of sense,
Get at essence, not with a rapier
But paper?

THE HAND

(For Helen Vendler)

The lady was sitting at a table at dinner,
Dignitaries on either side, conversation brimming,
It was an affair in the new-old academy, elegance
Hanging in the air like flowers of antique May.

The lady held her left hand moveless on the white linen
A very long time, well over an hour, a still life,
Remarkable in the fullness and planes of the flesh,
A rosy hand reminding of the bloom of Renoir,

Or it was immaculate as a day, perchance, with Rubens,
The flesh painted its own warm outreach of pleasure
Yet there was no motion, no sign of motion or emotion,
Presenting a vision of lightness, history, classical light.

The hand belonged to a living being, thriving now,
But gave no relation to the whole body, action, thought,
It had become immortal, a thing to itself alone,
A Platonic Type, eternity held in speechless accord.

THE MYSTICAL BEAST IN THE SHADOWS

I saw it, the mystical beast in the shadows,
Unbelievable, but I believed it
Because I saw it, but I did not see it,
The eye could not fashion its passing.

Sitting erect, in complete control of myself,
Thinking of nothing, aware of everything,
At ten o'clock of an autumnal evening
An animal appeared at the side of the house,

Moved toward me, unconscious of me, came
Within twenty feet in the light from the porch
Extending across the lawn to darkness,
The animal turned, and crossed my vision.

Too swift an event for determination,
This beast, perfected in semi-darkness,
Came into vision and so swiftly left it
That I was astonished, and remained astonished.

427

It was real, not of my imagination,
But darkness obscured its furtive shape,
I will never know what assailed me,
Was it the soul of time oppressing my space?

THE ANGELS

Some angels were standing on the ground
Unable to fly, their wings extended upward.

They kept this stance and pose year after year.
They were made of marble, unable to be human.

These creatures dazzled the ambient Florida sunshine,
Stood immaculate on the ground and never moved.

They looked like perfection, to be eyed askanse
Driving by year after year. Nobody would buy them.

They kept their residence near the lake,
A fisherman came by with big ruddy looks,

Went to the lake and took out a boat.
Green slime and blue water and white sky

Were all in motion when hook and mouth met,
The thrashing resplendent as the man hauled in

A big bright color of life for breakfast, and kept
Fishing all day in the belief of muscle and tone.

The white marble angels were always there, though,
Nobody would buy them, they could not move,

They were perfect and viewed life without expression.
When you passed them you looked but could not think a
    word.

I never forgot them they were so unexpected,
Breath-taking, out of this world, caught in it.

A decade does nothing to them. Nobody would
Buy them. Their marble wings never got off the ground.

They saw me age as I came by another year,
Took out a boat, went on the orange and weedy water

And caught the fish struggling to gasp for air
With no hope, giving his life for my breakfast,

Muscle and tone, while the angels were standing useless,
Unviolated, unable to fly, I thought to buy one

But what would I do with a stiff, marble, non–creature
So resolutely sub and super human, so final,

So far away from my desires and aspirations,
No suffering, no pain and joy, nothing for breakfast.

DEAD SKUNK

I buried a skunk in the garden,
He was beautiful, dead on the lawn.
Nobody knows whether a dog or poison
Left him there in the beginning of Spring.

I purposely made his grave shallow
To keep him as near to life as possible,
Just barely under the stuff of the earth.
We should not buy death too deeply

But keep it as if with us.
This was a beautiful of God's creatures,
Striped body, black and white,
A white, conclusive tail.

I put a rock over the place
As if to say resolutely
I honor your life, I mark your death,
And know little of either.

QUESTION MARK

Having thought all his life
He wanted to think
Something simple that would be universal.

All of his thinking had vanished,
He thought, with vanishing decades,
What was left of his thought?

What was the universal truth?
Could he put it in a simple word?
Could complexity be resolved?

Feelings had gone the way of thought
And now he was baffled
By the going away of feeling as well as thought.

He wouldn't give up to say death
Because he had life in him
But love and sleep tormented him

Also and all these big questions of time
Could never be answered,
So he wrote a poem.

Was the poem universally true?
Was it simple enough to be perfect?
It was simply lost

In the shuffle of time and thought
And lost feeling, death, love, and sleep.
The poem, the truth is, was never published

Because, the truth is, it was never written,
It had been written, as it were,
In a continuum of thought and feeling

Through a lifetime of found and lost
In a myriad of secrets kept by time,
Every breath he took was a question mark.

WAITING TO LEAN TO THE MASTER'S COMMAND

Three boats headed Northeast, moored,
Look like greyhounds straining at the leash.
They wish to go, but have no will of their own
Whether night or day, tide in or tide out.

These exquisite creatures have no imagination,
They cannot go to sea, to any harbor,
They have to wait on the command of man,
Man says go, they go, moving on Penobscot
    Bay.

THE BROKEN PEN

Will used a broken pen, must have broken many.
While he was writing Hamlet
And was in the midst of the deepest
Things he could say about life

431

The pen broke, sodden, and stopped,
This was the way life was
In the eldritch days of the Elizabethans.
Will had to make an instantaneous decision,

Better to lose the train of thought, let it go,
Difficult to make a supreme order
Out of the inveterate chaos of experience,
He started in truth, but the pen broke in the middle.

Almost impossible to think of William Shakespeare
As a man quite similar to the rest of us.
On the other hand, he was nothing if not human,
Must have been exasperated when his pen broke.

Our problems are not his problems. The atom bomb
Did not threaten his exquisite speculations.
William Shakespeare took another pen
And went on writing Hamlet to the end.

DEEP FISHING

Poetry is like fishing,
If you have six hooks
On a line one hundred feet down

What you have to do
Is wait for carp to strike,
A mystery of no feeling.

Haul up every half hour,
Often the bright beings
Are there, colorful catch.

You are way out in the ocean.
Percy is showing you how to fish.
In the distance is Egg Rock.

Whether strike or no strike
The ocean remains the same.
It is careless of you.

The ocean is the sea
Of creativity, dark down deep.
Memory is the line.

A caught fish is a poem
From the depths. Sometimes
They come, sometimes not.

The depths have made them virile
In their way. You are fishing,
Poetry is possible.

Fish die soon,
Poetry may live,
Ocean of imagination.

Percy starts up the motor.
If we go by Eagle Island light
We'll be in by dark.

Our boat was full of poems
That timeless summer day
So long ago.

THE DIFFICULTY OF IDEAS

Could he possess an idea?
They were like twittering birds
                 flying around.
Sitting in his obdurate flesh
Aware of the masquerade of time
He wanted poetry to be ethereal,
The age called for it to be material.

He wanted to break all the rules
To make a rule of his own—
Decades ticked away,
He could never discover his own rule.

Something about the twittering birds.
The ideas were always flying away,
Lithe and lively, remarkably pure,
His flesh was heavy and he had to laugh.

Could he possess an idea, idea the goddess?
The idea of a goddess was of the elusive,
Goddesses would not be possessed, but
Pursued as the idea of the unpossessable.

So the Greeks had sibyls supposedly human
Hidden in kinds of caves with only
                 a voice,
You did not see them, could not possess them
Yet they spoke immortal truths to
You amazed. Was it only an idea?
How real were these twittering personages?

His obdurate flesh called for
                       something more
Was it real? the inner? the outer? what
                    core, what for?

He had too many ideas, they
Were always flying away, vociferous birds
Trying to tell him the truth as they departed.
After his death they continued artful.

ON THE SUBTLE MAN

The subtle man, meaning the subtle woman too,
Was considered superior, but he was,
They were not known on the Square
Where traffic dealt in seasonal reality.

The obvious realities were the big
Realities of everyday necessity,
People were coming and going, doing or
Not, the grip of the world and life was fixed.

If you felt you had superiority
You did not want to mix with the crowd
Which could not understand your subtleties
Even if to them they were loudly avowed,

The subtle man was too good to be true.
He, or she, worked, perchance in the ethereal,
Whereas the others were locked in the material.
Materialists be damned, said the subtle man.

GULLED

It is said that man is gulled but I
Experienced the bit experience anew today
When a gull at lowest tide gulled me.

He also gulled Rick and Dr. Brown.
He seemed a young gull of this summer, dark
Not yet lighter, with one feather sticking out

Of one wing, which seemed excuse for not flying.
Rick and Dr. Brown were in kayaks, I afoot.
The gull would not take off. He paddled ashore.

Rick and Dr. Brown were paddling by the shore.
I walked down the shingle, daring to catch a bird in hand.
The bird walked off the seaweed, paddled by the bald
    rock

High out of water at low tide, which it now was.
I signalled to Rick and Dr. Brown to box him in
Against the high side of the steep egg rock.

We were three human beings, two in kayaks,
One walking, predators of a broken gull.
Something had been broken in the gull or he would go.

I really did not know what I wanted to do.
Did I want to catch this large wild bird in my hands?
Had we three any reason to interfere with nature?

It is important to assert that no malice
Existed in the hearts of three members of mankind
Curious about the inability of a gull to fly.

Curiosity, yes. Yes, a wish, somehow, how, to help.
None of the three knew the outcome of the drama.
If the wing broken, death the sooner the better, we thought

We thought. Nature exists in wordless impersonality.
Our little boxing-in operation began to fail
As the young gull began paddling out to sea
While the kayakers adjusted their oars and the landsman
Was still going out up to his knees. Then the sea gull,
With an abrupt indifference, an absolute disdain

Of man's concern for his welfare, raised his wings to the
     air,
Not maimed, not broken, not death-warranted,
Sailed out lowly over the bay in measured grace

And rose to the height of the other gulls of his society,
Flew with conclusions not to be foregone,
While all three of us felt gulled, each with different rea-
     sons.

CONFIGURATION

I am some man come out of some strange place,
My parents thought I was regular,
I was thought to unify these times,
Not to be daring, or spectacular.

As time went on I agreed
That I was a regular man, one of many,
But nobody saw the strength of my soul,
The desperate nature of my reality.

Therefore I was a modern dualist,
Outwardly serene, accepting the status quo,
Inwardly an endless questioner
Of why things are thus and so.

So with comedy I viewed history
And with passion viewed poetry,
From defeat to cultivate victory
From life's primal moiety.

This moiety is my talisman
And it is my advocate
To evoke timeless essence
From our timed sense.

VELVET ROCKS

A fern coming up shyly
Beside a stone
On a Spring day
In May

Pervades the silence
With
A deeper silence
Branching out.

If anything is perfect
The green fern
In May light
Presents itself.

Ramifying, still,
Open, air-touched,
Centuries
Grow delicate.

MEMORY

Vision like quanta shooting off any which way,
Paradoxical order of indeterminacy.
If the flesh would only let us go
Would the mind offer a satisfactory show?

The mind is flesh too, often mindless.
It comes and goes, solid, ephemeral.
Nature makes us live and die,
Obliterating the word eternity.

438

That girl in China once; the Bridge of Sighs, walking
Beside the Cam; reading Pascal in a hayfield in France;
Mt. Etna from the Greek theater in Taormina, waiting
For a goddess to appear in white, love without wisdom.

LAOCOÖN

Evil is such a great structure you can't surround it,
Can't contravert it, can't say it is not there,
An implacable essence in the middle of the road,
A block to the vision, static, holds off encounter,
You do not have any idea that you can overcome it
By intelligence, design, purpose, or practice,
It is a fact of nature in day and in night, whether
You are well or ill, old or young, favored or unfavored.

You have to carry on as if it were not there,
You cannot afford to be obsessed with its importance,
You have to ignore it, an agile thing to do,
To carry on your life in evenness, with even value,
You have to pretend that evil is not there, will not
Ruin you, does not live only for your destruction
If you want to do what you want to do with the day.

What you would want to do would be to be
The building sculptor with a marble theme, inflict man
With a massive, binding, twining serpent surrounding him
And show man with all his musculature, face
Clenched, holding off the serpent as if forever,
The conflict unresolved by fixation in sculpture,
Life going on for man and snake in enmity.

439

QUANTA

Heisenberg, Schrodinger, and Dirac
Your irrational atom jumps off the track,
The track does not know where it is going,
Indeterminacy our new method of knowing.

Newton said the apple fell by gravity
Regardless of man's virtues or depravity.
Now our sportive scientists insist on an atom
Jumping every which way back at 'em.

Mad as the idea of madness in man
This word says of anything "I can."

What paradox, the managers of the quanta
Can't manage them even if they wanta.

These same fellows see them jump like anything,
Wild, irrational, portentous, funny thing.

| The big | gang | shatter |
| No | | matter |

| | which | |
| | way | |

| Won't | | matta |

HORNETS BY THE SILL

There is something about a hornet
That wants to get out. Though inside,
A marvel and danger to the guests,
Hornets do not fly wildly around the room

But persist in looking through the glass
As if it were not there, constrained
By they know not what, consternated,
Contending judgmentally for freedom.

I am free to watch their lack of progress.
It is Autumn. Soon it will be dead cold.
Winter will have defeated the hornets.
They want to get out of this perturbative room.

They do not know why they are acting in vain,
Desperate articulation, buzzing as if in pain.

STATUE OF LIBERTY

I saw her as something in my inner eye,
Not as a statue standing in the sky.

Before there was a statue of liberty to show,
My ancestors came here long ago.

Every time I think of her I see
Where in our time we have come to be.

We forged across the growing land
And then we took a psychic stand.

We look up to her, her high-armed glory
As part of the American story.

She means more than what seems
Both in reality and in dreams.

Miss Liberty is tough and strong,
Promising her welcome will last long.

GOING

The comfortable craggy slope
You go up through knotted roots.
There is a long spell of walking
With little or no talking.

It is a long climb you are making,
A severe journey you are taking,
Under a canopy of twisted trees
Through which you can see the summer sea.

The sea is there in the distance,
You are aware of its existence
As you climb to the goal on the hill,
Something you can do if you will.

It has been done many times before,
Constricted road to an open door.
You brought what you had in your heart
Long before the arduous start.

Some came heavily, some came lightly,
Some have insight, some have no sight,
Some have nothing left to tell
After they have heard the warning bell.

A descent down the moss and rock face
Was necessary from that high place
Down into doubts, uncertainties
In a world of depths of mysteries.

NEW ENTREPRENEURIAL PLAN

If you can come to no conclusion about death whatever,
Direct your ashes to be whirled around in space forever.

442

Maybe to be placed in an orbiting capsule
On track will become the wormless general rule.

The entrepreneurs are selling lots in space, no commo-
　　tion,
Why not look down forever on the human race, in mo-
　　tion?

Whether life ends in fire or ice
Your social security number is included in the price.

21ST CENTURY MAN

Finally, he decided there was too much pain,
The hurt of everything.
In youth it was not knowing,

In middle age it was knowing,
In age it was not knowing.
He couldn't figure it out.

Would 21st century man do better
Or 21st century woman do better either?
The tides were always going in or out,

But what was the meaning of the ocean?
People were either growing up or growing down.
He decided to live for sensual reality,

Pure feeling. After this failed
He decided to espouse pure intelligence.
This never told him why he had to die.

He then decided to go to the Church
But after the supreme fiction of Christ
He thought Buddha and Mohamet had something to say.

Neither sense, intellect, nor religion
Told him why he was born or had to die
So he began to pay attention to poetry.

Non-suicidal, he desired to make something.
He decided the greatest thing was a perfect poem.
If he could make it he would be glad to live.

The brutal fact, dear reader, as you
Might suspect, is that he did not make it.
Somebody else made his perfect poem imperfect.